Measuring Food Sustainability
and the Benefits of
Urban Agriculture

Dr. John Zahina-Ramos

Library of Congress Control Number: 2018906596

ISBN-10: 0-9863795-2-2
ISBN-13: 978-0-9863795-2-9

The author or publisher is not responsible for websites (or their content) that are not owned by the author or publisher.

For more information or updates concerning this book, visit the Just One Backyard website at www.justonebackyard.com or contact the author at sustainable@JustOneBackyard.com

Cover photo: German homestead in Old World Wisconsin, photo by John Zahina-Ramos

Table of Contents

List of Figures.. ix

List of Tables xi

Foreword xv

Preface...xvii

Chapter 1. Introduction 1
Purpose of this Book... 2
   Objective Metrics 3
   Food Sustainability Definition and Quantification 3
   The Residential Food Garden Case Study.............................3
State of the Science 4
Judgment, Morals, Ethics and Science 4

Chapter 2. Why Quantify? 9
Advancement of Agricultural Knowledge 9
Measure Benefits and Harm from Agricultural Systems............10
Inform Planning and Policy Decisions.....................................13
Support Sustainability Initiatives..14

Chapter 3. An Overview of Agriculture 17
Commercial and Non-Commercial Agriculture..........................19
   Commercial...20
   Non-Commercial 21
Production Methods...21
   Traditional 22
   Organic 24
   Industrial 25
      Hydroponic..26
      Container...26
Crops...27
   Traditional crops 27
   Specialty Crops 27
   Subsistence Crops...27
   Crop & Livestock 28
   Unusual Crops 28
Scale and Settings..28
   Large-Scale Commercial Farms 29
   Small-Scale Farms..29
   Community Gardens 29

Home Food Gardens 29

Chapter 4. Sustainable Urban Agriculture 31

Urban 32

    The Physical Environment 34

    The Social Environment 35

Sustainability 37

Sustainable Agriculture 39

    Objective 1: Meet Present and Future Needs 40

    Objective 2: Resource Stewardship 41

    Objective 3: Environmental Protection 42

Chapter 5. Environmental Conditions and Agriculture 45

Plant Productivity 45

Climate 47

    Growing Season Duration 48

    Climate and Latitude 49

    Climate and Elevation 49

    Temperatures 50

        Average Temperature 51

        Temperature range 51

        Temperature Extremes 52

    Wind 52

Moisture Availability 52

Soil 54

Agricultural Pests 56

Complex Landscapes 56

    Urban Development Patterns and Forms 57

    Ecological Gradients and Patterns in Cities 57

    Environmental Gradients and Resources 59

Chapter 6. The Economics of Urban Food Production 61

Food Growing Operation Size 62

Costs and Expenses 63

Commercial Farms 64

Non-Commercial Food Growing Operations 67

    Farms 67

        Subsistence Farms 67

        Non-Profit Farms 68

    Community Gardens 69

    Residential Food Gardens 70

    Other Food Growing Operations 70

Chapter 7. Quantifying Agricultural Inputs 71

Tangible Inputs to Food Growing Operation 72

Planning 73
Crop Selection 73
Preparation: Materials & Infrastructure 74
    Soil Preparation 76
    Soil-Like Medium Preparation 76
    Non-Soil System Preparations 76
Planting 76
Growth 77
    Water 77
    Light 77
    Nutrients 77
Care 78
    Monitoring 78
    Pest Control 78
    Protective Measures 79
Harvest 79
Intangible Costs of Food Growing Operations 79
Quantifying Food Production Inputs 80
Direct and Indirect Costs 80
Materials and Supplies 81
Labor 81
Monetary Investment 84
Water Usage 85

Chapter 8. Quantifying Agricultural Returns 87
Measuring Tangible Returns 88
Defining the Crop 88
Measuring and Expressing Productivity 89
Return Valuation 92
Quantifying Intangible Returns 94
Identifying Intangible Returns 94
    Psychological 94
        Personal Enrichment and Well-Being 94
        Cultural Expression and Connection with Heritage 96
    Food and Nutrition 97
    Community 98
Measuring Intangible Benefits 99
Indirect Costs and Benefits of Urban Food Growing Operations 100

Chapter 9. Environmental Resources and Urban Ecology 103
Environmental Resources 103
Water Resource Use and Conservation 104
Energy Resource Use and Conservation 107
Waste Product Generation and Assimilation 111
Plant Nutrient Use, Conservation and Management 113

Other Resources 116
Ecological Resources.. 116
   Quantifying Ecological Aspects of Urban Agriculture 118

Chapter 10. Cost-Benefit Analysis for Urban Agricultural Systems **121**
The Cost-Benefit Analysis: An Overview 122
The CBA Framework 124
   The CBA Setup 124
   The CBA Structure and Calculations... 125
   Valuing Hard-To-Measure Effects 128
Issues with the CBA 129
   Exclusion Bias.. 129
   Risk Equity 129
   Valuation Theory 130
   Cost-Effectiveness vs. Cost-Benefit Analysis................................... 131
The CBA and Urban Agriculture... 132
   Time Valuation... 132
   Growing Methods and Materials 133
   Purpose of the Food Growing Operation 133
   Water... 133
   Food Valuation 134
   Energy and Greenhouse Gas Emissions 134

Chapter 11. Quantifying Food Sustainability **137**
What is Food Sustainability?.. 137
Quantifying Food Production... 143
   The Foodshed 143
   The "Healthy" Diet... 146
   Quantifying Historic Food Production (FP_H) 146
     Agricultural Census Data and Surveys 148
     Crop Production Studies... 149
     Commercial Agricultural Crop Data 149
     Proxy Data 149
     Other Crop Data 150
   Quantifying Potential Food Production (FP_P) 150
Quantifying Potential Food Demand (FD_P)..................................... 151
The Food Sustainability Measure (FS_M) 153
The Food Sustainability Index (FS_I) 154

Chapter 12. Residential Food Growing Operation Case Study- Methods ...**157**
Urban Food Garden Study's Goals and Objectives 158
   How Much can be Produced?... 159
   What Are the Costs and Benefits?... 160
     Environmental and Ecological Costs and Benefits...................... 160
     Economic Costs and Benefits.. 161

Social and Intangible Costs and Benefits 161
Can Enough be Grown to Meet Residents' Needs? 161
Urban Agriculture and Food Sustainability 162
Study Site 162
Study Duration 167
Food Growing Site 169
Data Collection Methods 170
Environmental Parameters 170
Water 170
Energy 171
Waste 173
Fertilizers 175
Productivity 175
Harvest Valuation 176
Labor 177
Expenditures 177
Ecological Parameters 177
Social Parameters 178
Potential Residential Food Production in the Metropolitan Area 179
Estimating Available Residential Food Growing Space 180
Estimating Food Production Potential (FP$_P$) in the Metropolitan Area 183
Estimating Potential Food Demand (FD$_P$) of the Metropolitan Area 183

Chapter 13. Residential Food Growing Operation Case Study- Results **185**
Climatic Conditions 185
Food Garden Productivity 187
Value of Food Grown 187
Production Costs 190
Ecological Aspects of the Food Growing Site 193
Local Greenspace 193
Uncultivated Species Observed 194
Environmental Costs and Benefits 199
Waste Generation and Assimilation 201
Costs Associated with Energy Consumption 202
Intangible Costs and Benefits of Residential Food Gardens 205

**Chapter 14. Residential Food Growing Operation Case Study- Sustainability
Analysis** **209**
CBA Assumptions 210
Labor 210
Energy Consumption and Conversion 212
Environmental Costs and Benefits 213
Water Resources in the Local Context 213
Waste Products 214
Impacts to Environmental and Natural Resources 216

Urban Ecology 217
Examination of Costs and Benefits of the Residential Food Growing Operation...... 219
Monetary-Based Project CBA 219
Comprehensive Project CBA 221
Comprehensive Efficiency CBA 223
With-Project/Without-Project Analysis 226
Sensitivity Analysis and Uncertainty Management 227
Residential Food Production and Food Sustainability.............................. 229
Household Food Sustainability Analysis 230
Community Food Production and Food Sustainability 232

Chapter 15. Towards a Verifiably Sustainable Agricultural System 237
When Sustainable Agriculture Science meets Culture, Policy and Politics.............. 237
Culture 238
Policy 239
Politics 242
The Pendulum and Patience... 243

Glossary and Abbreviations 245

Index 257

References 269

About the Author 283

List of Figures

Figure 1. Conceptualized relationships between environmental, economic and social sectors of sustainability...38

Figure 2. Generalized climatic zones associated with latitudinal changes in the Americas 48

Figure 3. Representation of the complex urban environment and extreme heterogeneity of growing conditions. 58

Figure 4. Digital aerial photo of the Miami, Florida metropolitan area c. 1984 showing a gradient of built, planted and natural land cover 58

Figure 5. Inputs required by urban food growing activities...71

Figure 6. Tangible and intangible returns realized (i.e., effects caused) by urban food growing operations..88

Figure 7. Solid waste costs per population center size in North Carolina, 1994-1996.. 112

Figure 8. Sample economic cost-benefit analysis structure. 126

Figure 9. Food sustainability index value relative to the population's sustainability condition and food supply. 155

Figure 10. Map of the Palm Beach County metropolitan area in south Florida, USA 164

Figure 11. Location of the food growing study site within an urban residential neighborhood 165

Figure 12. Extent of agricultural lands in Palm Beach County, Florida in 1988 and 2009 166

Figure 13. Map of the residential food growing study site showing production bed locations and the adjacent wildlife garden. 170

Figure 14. Residential urban development in metropolitan West Palm Beach, Palm Beach County, Florida, USA 179

Figure 15. Township-Range-Section blocks in Palm Beach County indicating randomly-selected residential blocks used in this study 181

Figure 16. Sample residential Section and SW quarter-Section tract in metropolitan Palm Beach County, Florida, USA. 182

Figure 17. Sample aerial photo of a mixed multi-family and single-family residential community in metropolitan Palm Beach County, Florida, USA. with potential food growing areas indicated. 183

Figure 18. Digitally processed image visualizing greenspace within the study area neighborhood 194

Figure 19. Association between a history of growing food within questionnaire respondents' families and the likelihood respondents were food growers 206

Figure 20. Questionnaire responses related to why local residents grew food for themselves... 206

Figure 21. Intangible benefits reported from home food growers 207

Figure 22. Monetary-based (internal effects) project CBA for the residential food growing operation 220

Figure 23. Comprehensive project CBA for the residential food growing operation 222

Figure 24. Comprehensive Efficiency CBA for the residential food growing operation 224

Figure 25. With-Project CBA for the residential food growing operation............ 228

Figure 26. Questionnaire respondents' feelings about residential food growing 240

List of Tables

Table 1. Estimated world population from 1800 - 2050 12

Table 2. Elements typically included in sustainability initiatives. 16

Table 3. Examples of agricultural products. 17

Table 4. Average growing season durations for selected cities 50

Table 5. Temperature statistics that are important to food growing operations 51

Table 6. On-going expenses for urban agricultural operations 64

Table 7. Time/labor investment parameters 82

Table 8. On going and seasonal inputs to food growing operations 83

Table 9. Estimated average labor needed to grow selected crop types 83

Table 10. Typical expenditures of urban food growing operations 84

Table 11. Water sources typically available for urban food production 85

Table 12. USDA's Choose My Plate recommendations for fruit and vegetable consumption 92

Table 13. Intangible benefits associated with non-commercial urban agriculture 95

Table 14. Indirect effects associated with urban agriculture operations 101

Table 15. Monetized damages per unit of energy-related activity 110

Table 16. Vegetable varieties within Choose My Plate vegetable subgroups 147

Table 17. Modified Choose My Plate age categories and recommended weekly consumption of fruit and vegetables 153

Table 18. Factors measured in the food growing study 163

Table 19. Palm Beach County population from 1970 through 2014. 165

Table 20. Climate data for West Palm Beach 168

Table 21. Data collection periods in the study. 168

Table 22. Space partitioning during the backyard food production study 170

Table 23. U.S. Census demographic data for Palm Beach County 184

Table 24. Environmental conditions during the case study 186

Table 25. Types of vegetables and fruits harvested from the residential
food growing study over the study period 188

Table 26. Total weight of produce harvested from the residential food
growing study during each growing season 189

Table 27. Value (2016 USD) of harvested produce from the residential
food garden over the study period 190

Table 28. Total annual water usage during the case study 191

Table 29. Time (labor costs by hour) spent on the residential food
growing operation. 192

Table 30. Monetary investment during the case study 193

Table 31. Uncultivated plant species that were observed at the
residential food-growing site. 195

Table 32. Invertebrate species observed at the food-growing site 196

Table 33. Agricultural pest predators found within the residential food
garden study's region 197

Table 34. Pollinators found within the study site's region that are
associated with crop types grown in the research garden plot 197

Table 35. Vertebrate species observed at the food-growing site 198

Table 36. Indirect costs of the residential food growing operation 200

Table 37. Non-recyclable, non-reusable waste generated by the food
growing operation 201

Table 38. Organic material that was composted and added to the
residential food garden 202

Table 39. Environmental costs related to electricity and fossil fuel
consumed during the food growing study 203

Table 40. Climate-related impacts that would have occurred if
grown produce had been purchased from local grocery stores 203

List of Tables

Table 41. CBA elements (effects of the food growing operation), grouped by analysis type 210

Table 42. Labor valuation based on garden productivity 211

Table 43. Known indirect costs that were identified from materials used or consumed during the residential food growing operation 218

Table 44. Number of servings of fruit and 5 categories of vegetables harvested during the residential food garden study 231

Table 45. FP_P for metropolitan West Palm Beach, Florida, United States. 233

Table 46. Palm Beach County Metropolitan Area's FD_P 234

Table 47. FS_M and FS_I values for different food types that would results from growing produce in existing growing space in residential backyards 235

Foreword

We humans eat every day of our lives, from birth to death, three meals per day and between meals as well. Our food intake supports our physical, social and spiritual being. We are absolutely unsustainable without food. What could be more basic to our lives than the act of eating? And, specifically, what could be more economically important to our lives and to our society than the choices we make concerning what to eat and how much to eat? Food, and our decisions and choices around food, are power, in fact, ultimate power. And we know also that eating is a moral act, that is, an act with moral consequences. Every bit of food we purchase and consume registers in the economy, whether from retail grocer, restaurateur, or fast food take-away. Thus, each food purchase sends a signal to grow, raise, process, prepare and distribute, for there is a guaranteed sale at that price at the retail end. And money is borrowed and invested, based on those figures. So, food counts! And food is a big way for the individual, for any of us, to have an impact on the national and local economy. That impact comes through our avoidance of participation in the national economy and its large-scale industrial system of agriculture, of factory farming. And it comes through participation in, and through support of, local food and farming, thereby building and strengthening the local economy and, importantly, insuring the kind of local food security which could never be possible with the current national and international system of food production and distribution. This is where John Zahina-Ramos' book, Measuring Food Sustainability and the Benefits of Urban Agriculture, enters the picture.

John is an experienced expert on the power of home gardens and local food production, and this volume illustrates that expertise. The author writes as an educator and an academic meeting the demands of scientific and economic research. He thus has a good handle on the quantity of food that a well-designed and well-managed garden can actually produce, and under a variety of circumstances, using the best of available small-scale modern technology and the latest ideas. But, as an experienced practitioner himself, he also understands the realities faced by the amateur gardener, the average home gardener.

It's reasonable to expect that home (and all local) garden production will continue to increase; that season extension technology (hoop houses, high tunnels, grow tunnels), based on renewable energy, will continue to proliferate; that sustainability practices like greater crop diversification (even including grains), and companion planting to fight insects and disease, will be utilized; and that animal inputs will also be utilized (poultry, perhaps rabbits, pigs, goats) to both provide economic diversification and particularly to provide needed soil fertility for crops.

The result of all of this activity will be a proliferation of healthy local food (and thus an increase in human health), a stronger local economy, a stronger community, more good and rewarding local jobs, and, in some ways most important, food security in every local place. For all of that, we will have John Zahina-Ramos and other like-minded souls to thank. So, read this book and thank its author! Above all, use your power to secure and control your own food system. Live free and farm – or garden! And support your neighbors in doing so. You and your country will be well served.

* * * * * * * * * * *

John E. Carroll, Professor-Emeritus of Environmental Conservation, University of New Hampshire, is author of *The Wisdom of Small Farms and Local Food*, *Pastures of Plenty*, *The Real Dirt* and, most recently, *Live Free and Farm: Food and Independence in the Granite State*. He lives in Durham, New Hampshire.

Preface

This book is the culmination of 10 years of inspiration, vision, hope, determination, learning, planning, implementation, learning, data collection, data analysis, learning, review, re-thinking, learning, editing, ups, downs, learning and, eventually, success. That is the true step-by-step process of every science research project. What happens after the work is done is perhaps the most important part of all. A researcher can formulate a sound hypothesis, collect the finest data, conduct appropriate analysis, arrive to undeniably important conclusions and write the most persuasive documentation- but if it is not read, understood and integrated into the fabric of broader knowledge, the effort will be pointless. The scientific literature has become an unwieldy bulk of such material and is increasingly impossible to synthesize, even by the experts themselves. The quick pace that scientific knowledge is advancing has also caused the practical side to be left behind. For example, atmospheric scientists can accurately describe the chemical basis of climate change at the molecular level, but many citizens and community leaders question whether climate change is real. The problem is not our ignorance about the mechanisms of climate change, it is that scientists have spent so little time explaining the relevance of their research to audiences beyond their academic bubble. Merely publishing the results of one's research in science journals is not enough. It must also reach the audiences that can benefit from it or are impacted by it.

For this reason, I have decided to not pursue the traditional route of publishing this work in science journals. To do so would have required splintering the research into discreet pieces in order to fit into the confines of word count restrictions and subject matter focus. Agriculture, by nature, is multi-disciplinary. So is sustainability. Then, combine these two fields of knowledge within the context of the urban environment…the result is a subject that requires a broader discussion than can fit into a single, or several, journal articles. The complete picture of the relevance, methods, rationales and results of this research are best offered through a larger work; that is the purpose of this book. Besides scientists, this study and its results must also be available to practitioners of urban agriculture, community leaders and decision makers…otherwise it cannot effect change. To reach this latter audience, a more open and affordable approach to access is necessary.

This book has been written with an academic voice and is suitable as a textbook on sustainable agriculture, urban agriculture, agricultural sustainability, urban sustainability, food systems and quantitative sustainability. It is also a methods book that community leaders can use to calculate the sustainability benefits of an existing or proposed project. This book is counterpoint to its predecessor, a book that I published in 2015 entitled *Just One Backyard: One Man's Search for Food Sustainability*. This latter book was written for the general audience who has no background in science. It explains the same concepts and conclusions produced by this research, but in words the average person can appreciate.

In the spirit of open-access and open-review, all of the data collected during this study is freely available for download, review and analysis. It can be obtained from my website: www.JustOneBackyard.com. If you have trouble finding the data, send me an email and I will provide you with a copy. Secondly, in the spirit of sustainability, I have kept the printing of this particular paperback edition limited to black and white (with the exception of the cover). Although it is normal to create books that pop with color and use eye-catching graphics (which I may offer in other editions), they also consume resources that are not essential to the printing of this work. It seemed wrong to publish a book on sustainability while not employing more sustainable methods for printing the book itself. By using a minimum of color inks, the paperback version of this book can be completely composted except for the color cover pages, which can only be recycled. It also makes this work affordable to those who would otherwise not be able to acquire a copy. Lastly, this research has been self-funded; no organizations contributed to this research financially or otherwise.

I wish to thank my late aunt Liz and uncle John for being the inspiration for this unlikely study. I sought to replicate her Old World kitchen garden in an effort to understand not just its benefits, but what has been lost by society's transition away from local food growing. Lastly, I want to thank my spouse, Eddie Ramos-Zahina, for bearing with the thousands of hours of work I invested this effort, particularly since he dislikes vegetables. Certainly, we are both happy it is finished. ☺

All the best!

Dr. Z
Chicagoland
May 25, 2018
sustainable@JustOneBackyard.com

Chapter 1. Introduction

Food. Along with oxygen and water, food is among the most fundamental requirements for human life. The food that people eat must be clean, varied and certain amounts must be taken in every day to maintain health. The way this food has been grown can have consequences well beyond the farm field and practices that pollute or consume nonrenewable resources will eventually degrade the fields themselves and other lands far from the growing operation. Additionally, what happens between the farm field and the consumer can have economic, social and environmental ramifications that can be unpleasant, at best. Given the complexity, great importance and enormous scope of the modern food system, studying it can be challenging.

Just because a system is complex, does not mean that it cannot be adequately studied. Biologists classify species within the Amazon rainforest despite its vastness. National and international space exploration programs send probes and sensitive monitoring equipment to places that have never been visited before and that are unimaginable distances from Earth. Physicists continue to break knowledge barriers by looking at ever-smaller particles and their behavior. From these examples, it is clear that complexity alone is no reason to avoid inquiry.

The ways that food production methods impact the environment are complex. The meaning of food to people is complex. The reasons that some people grow food (and others don't) are complex. Food systems are not just complex, but ever changing. The key to studying complex systems is to look at the individual parts, understand how they work, and then look at how these parts interact. It is that approach that this book is founded upon.

When one calculates the amount of food that is consumed, or should be consumed, by an individual over the period of a year, it is astonishing. When calculating this volume of food for all people living in a large metropolitan area, the amount is beyond comprehension. Going through the exercise of calculating this volume of food leaves one with an appreciation for the role that agricultural systems play in maintaining human life. It also begs the question of where all that food comes from, how it was grown and what that might mean.

Calculating the volume of food that is necessary to feed the residents of a particular city is only part of the story. It is also essential to know how that food was grown, where it was grown, how much space was needed to grow that food, what resources were consumed to grow it, how far that food traveled from the point of production to the point of consumption and what negative consequences may have been generated by producing that food. As the amount of food demand generated by a growing global population increases, so do the costs. Most consumers are unaware of or not concerned with these costs, except when it affects the money they have to spend to purchase it. But, one way or another, the array of costs mentioned above ultimately does affect humanity.

The quest for a sustainable agricultural food system is grounded in the desire to reduce the costs of producing humanity's food, particularly at the environmental level. The world's population has grown exponentially since World War II and does not appear to be slowing. The amount of resources needed to grow more and more food has caused natural resources to be consumed in unsustainable ways and the extent of natural lands has shrunk to the point where many species have been driven to extinction because of habitat loss.

Purpose of this Book

This book examines urban agricultural systems in new ways- by demonstrating the value of quantitative metrics to estimate potential production, by pointing out new ways that agricultural production should be viewed and valued, and by showing how urban agriculture may be sensibly applied to address social and economic deficiencies through an informed process.

This book has been written with several purposes in mind. The first is to present objective metrics for measuring the costs and benefits of urban agriculture, which can take on a variety of forms and functions. Once these parameters have been measured, they can be examined to ascertain just how beneficial a food growing operation (or method) is. They can also be used to compare between different growing methods or operations, or to run an analysis of different growing design scenarios to determine which is a better choice. The second purpose of this book is to present a functional definition of food sustainability and to demonstrate how that can be quantitatively measured. Lastly, this book presents a case study that quantitatively measured the environmental, ecological and economic costs and benefits of a residential food garden. This case study is used as an example of how to apply some of the quantification methods discussed above and as a contribution to a poorly studied area of the agricultural sciences.

This book is suitable for the following users and applications:

- Practitioners of urban agriculture to better understand how costs, expenses, inputs, benefits and outputs can be objectively measured. These users can also use the methods in this book to calculate the suite of costs and benefits resulting from different growing scenarios for the purpose of selecting an optimal plan.

- Planners and policymakers who are involved in scoping new urban agriculture initiatives or projects or who wish to identify the benefits of existing projects; this book can guide data collection and analysis, as well as provide the context that the results should be placed in.
- College courses in sustainable agriculture, agricultural systems or food systems.

Objective Metrics

The field of sustainability sciences, which is a relatively new field of study, typically relies on qualitative descriptions of practices and processes rather than a body of scientific evidence that has been obtained from detailed studies. Although sustainability is often talked about in relative terms (i.e., one practice is relatively more sustainable than another), it is often hard to define in an absolute sense. As the field of sustainability sciences matures, it will demand that objective and quantitative metrics, which yield direct evidence, be used to determine if a practice is sustainable.

Food Sustainability Definition and Quantification

A review of the scientific literature yields an anemic and inadequate definition for food sustainability, regardless of source. This is unfortunate. There has never been a higher demand for a sustainable food system, yet it is currently not possible to know if one attains food sustainability or how it can be achieved. A measurable, achievable and practical definition is needed; if it isn't, it will not be able to be widely understood or calculated by those who most need it.

The Residential Food Garden Case Study

It is one thing to present a suite of analytic methods, but without demonstrating how these can be carried out, interpreted and synthesized with other information, the concepts can become cloudy and detached from context. The case study was conducted in an authentic residential food garden that was located in metropolitan West Palm Beach, Florida. This case study is different from so many other agricultural production studies in that it has been carried out in a real-world setting (not under the ideal conditions of a field test plot) and data were collected on a comprehensive suite of environmental, ecological and economic factors that are common to urban agricultural operations anywhere.

Case studies have inherent limitations and these must be fully respected. However, these studies are valuable for examining how processes work, the relative importance of some factors and the results may be informative for other sites. The methods used to quantify costs and benefits may need to be adapted to be applied in other regions, but are generally applicable everywhere.

State of the Science

The subject of sustainability has become increasingly popular since the beginning of the 21st century. Scientists have struggled to qualitatively describe, and more recently quantitatively define, what sustainability is and what it means. Presently, little quantitative methodology exists to measure sustainability in ways that allows meaningful comparisons between practices to ascertain which is more sustainable. Through time, sustainability concepts and metrics will mature. But this process cannot happen without discussions across many academic disciplines and will be impotent if the scientific community does not build bridges to popular culture, ordinary people, policy makers and politicians. The enormous gulf that currently exists between mainstream society and the scientific community is proof of this. The scientific literature, particularly scientific journals, is a vast and unwieldy collection of detailed data, findings and facts that are linguistically, intellectually and functionally beyond the reach of most people. Until intermediates are put into place that bridge the gap between academics and popular culture, the differences between the two groups will only widen and sustainability will remain relegated to the extremes of scientific exercise and product marketing hype.

Many adherents to food growing movements (e.g., organic, permaculture, etc.) and practitioners of urban agriculture label their efforts as "sustainable" because they believe they are a better alternative than food produced using industrialized production. For the most part, those who produce food using a process that is considered better than the worst-case scenario often describe their practice as "sustainable". Without specific ways to measure how sustainable a practice is, based on appropriate data collection and analysis, these labels may be inaccurate or patently false, however well intended. There currently exists few guidelines or methods to quantitatively measure the degree to which a practice is sustainable. This book is intended to contribute to the conversation about how food production and food sustainability can be measured, and to present novels ways to do so. It is hoped that these methods will fill an existing void in the sustainability sciences and will place uncomplicated and straightforward methods into the hands of practitioners and analysts. In doing so, the seemingly vague and mystical world of food sustainability will be brought into the spotlight of quantitative data analysis and critique.

Judgment, Morals, Ethics and Science

Scientific studies, scientific knowledge and the science community are human constructs and, as such, are subject to the peculiarities of human behavior. The peculiarities that are most bothersome to quantitative scientists are the uncertainties brought about by the irregularities of human behavior and the role of morals and ethics in the interpretation and application of study findings. Irregularities can include perspective bias, inconsistent interpretation, mistakes, misunderstandings, etc. However, with these negative human traits also come other positive ones such as rational thinking, creativity, pattern recognition, learning from past experiences and awareness of threats; these can be beneficial to scientific research and can place findings into contexts that can rank the importance of something relative to a certain objective. To further complicate matters, these traits are applied in a selective or unpredictable fashion. The irregularities of human behavior are often difficult or impossible to fully separate from the scientist. At best, irregularities that may influence study conclusions are managed or avoided through controlled data collection methods, appropriate data analysis and independent peer review.

Scientists can be divided into three camps relative to the issue of if or how morals or ethics should be involved in the process of data analysis, the interpretation of results (e.g., the context that the results are placed in) and how the results are applied. These can broadly be defined as the empiricists, the moralists and those who fall somewhere in between. This issue has special importance in the arena of urban agriculture because of how integrated the practice is with social and economic aspects of the community.

The pure empiricist will argue that judgment based on human values (that leads to conclusions) and morality (or ethics) has no place in the reporting of scientific findings; this comes from a concern that human values and ethics will bias the reporting and application of scientific findings. The pure empiricist will provide data and report findings to colleagues and students in a clinical context, leaving it to the hearer to determine what this information means and how it should be used. This is a very democratic perspective that carries an assumption that the hearer has adequate critical thinking skills and understands the context that the information should be framed in. However, when the hearer is not astutely proficient in these skills, the information immediately acquires bias when it is filtered through the listener's perspective. Scientists who hold to this purer form of empiricism relieve themselves of the responsibility of interpreting findings in a way that may later be challenged or found to be incorrect. However, the risk of having one's interpretation or conclusions challenged is at the very core of scientific research. By its own nature, research and the scientific method are designed to question, challenge and verify the veracity of scientific judgments and conclusions.

The pure moralist holds the perspective that it is the responsibility of scientists and academics to make judgments about what data means and how it should be applied. For these scientists, the reporting of sterile facts and figures is not enough. For the moralist, data should be analyzed empirically and be interpreted through a context of values, ethics and morals. It is only in this context that students and other scientists can understand what the data means, why it is important and how it would best be applied. The potential drawback of the moralist approach is that the individual values of the investigator could be infused into the interpretation, rendering different interpretations from different investigators using the same study data.

One difference between the empiricist and the moralist perspectives can be illustrated by the following example. There is a well-studied relationship between species abundance and the availability of essential resources, particularly with respect to the concept of a carrying capacity. The carrying capacity of a resource limits a species' population size in an area. The human species can use a limited essential resource indefinitely so long as its use is not greater than the rate at which the resource is replenished. But, if humans use the resource in an unsustainable manner by demanding more than can be replenished, the resource will become exhausted and humanity will suffer significant consequences. Both the empiricist and moralist will report these facts. However, the moralist will point out that scientists also have an ethical obligation to warn society of this danger and to encourage a change in resource use for the good of humanity and the environment.

In reality, most scientists fall somewhere between the two poles of empiricism and moralism, depending on the situation and subject. One needs to take only a cursory review of the scientific literature relative to one specific topic to understand that much of the scientific community behaves in this manner, regardless of whether they profess to be an empiricist or moralist. When considering the subject of this book, which is rooted in two topics (agriculture and sustainability) that are of paramount importance to humans, it would be impossible to lean exclusively on the side of empiricism. In fact, the main concern of this book is to define ways to measure how sustainable food growing practices are and to use that information to help humanity move towards agriculture that will be productive over very long periods of time without contributing to environmental or social decline. After a study has been conducted and the results tabulated, it is necessary to judge whether one practice is more sustainable, and therefore, more desirable than another. In deference to the example of carrying capacity mentioned in the previous paragraph, the endeavor to measure sustainability carries with it an inherent moral imperative and judgment between options for the sake of protecting the environment and increasing humanity's chances for survival well into the future. The application of the methods described in this book will yield data and information that should not just be reported, it should also be interpreted according to its context and considered for what it could mean to humanity and the health of the world's ecosystems.

At this point in human history, the world's population has never been so high, the extent of urbanized land has never been so great, the demand for food has never been so enormous and the loss of natural habitat to human settlement so profound. Rural, industrialized agriculture was developed to feed the growing human population of the 20th century. That agricultural system has been a primary contributor to natural habitat loss and degradation; it is an unsustainable model and cannot continue to be the world's major source of food. Humanity is facing an uncomfortable dilemma- the negative impacts from industrialized agriculture is both feeding the world and slowly eroding mankind's ability to survive on this planet. This dilemma is not a mental exercise of logic or fodder for a varsity debate club. One can easily argue that it is also a moral issue, especially when so much food is being produced and so many people are starving or have poor nutrition, when a commercial market system has made food an item only within the grasp of those who can afford it, in that each acre of agricultural expansion destroys another acre of the environmental goods and services necessary to support human life. It is both illogical and immoral to continue to rely on such a food system to meet the needs of billions of people who will be born over the next decades when it cannot meet today's needs nor can it protect the environment that human life depends upon. From this perspective, enters the moralist.

Chapter 2. Why Quantify?

Why quantify the benefits from agricultural production? For growers, there is an obvious interest in knowing how much had been produced during the previous season and what that means from an economic perspective. This is particularly true for commercial operations. If one wishes to run a profitable farm, it is essential to know how much was grown (or could be grown), what the value of the crop is (or potentially could be) and what is required to make the operation economically viable. Unfortunately, knowledge about the functions and benefits of urban agricultural systems is scant, leaving a void that academia has left mostly to the qualitative sciences. But for policy makers, planners and politicians who need to make decisions about what the future of local agriculture will look like and how it can address social issues, the need for data and evidence to support any particular decision is critical. For them, the profitability of a food growing operation is often of lesser importance than other benefits the activity produces. Another vital consideration is the protection of the environment. Agriculture often comes with environmental and ecological costs that must be known and weighed against the good that the food growing effort can yield. These are some of the reasons why the benefits and costs of urban agricultural systems must be quantified and shared freely amongst stakeholders and the public. Below is more discussion about these issues.

Advancement of Agricultural Knowledge

Agricultural knowledge can be separated into different, but often overlapping, areas of expertise and focus. The depth and breadth of knowledge within a given area varies and is not the same for commercial and non-commercial agriculture. Agricultural universities have invested an enormous amount of resources into researching commercial industrialized production methods, quantifying a wide range of factors that are involved with production. These include: the number of seeds per row length, row spacing, number of days to harvest, the amount of specific plant nutrients that should be applied per unit area and expected yield for a specific crop plant. There is also a rich body of knowledge about commercial farm economics. An underlying belief of most modern agricultural research is that production intensification is a desirable goal. However, as agriculture becomes more intensive in practice, the ecological, environmental and societal costs also increase. As these costs mount, there is a point where the added benefits of incremental agricultural intensification become too small to justify. Much of the agricultural research conducted today is focused on advancing industrialized agriculture. In contrast, relatively little is known quantitatively about the productivity, benefits and economics of urban and non-commercial agriculture.

Over the past decades, advances in technology and data collection permit a more robust assessment of food growing methods and one can conduct a "life-cycle assessment" of costs and benefits of an activity. The life-cycle assessment (sometimes referred to as the "cradle-to-grave" analysis) examines all of the costs and benefits of all of the inputs and outputs of an operation throughout all stages. It is now possible to characterize and measure the resources that are consumed during the manufacture of materials used to grow food. It is also possible to measure many of the intangible benefits that result from the agricultural operation and the importance of these to the community. But, to do so requires one to call upon both the hard (quantitative) sciences and the social sciences.

In addition to costs and benefits, other questions of importance to urban and non-commercial agriculture are:

- How feasible is a proposed venture?
- Is there enough space to create a productive operation?
- What is the production potential?
- Can a proposed venture contribute significantly to the local food supply? If so, how much?
- Will people in an urban area grow food for themselves? Are they receptive to the practice and if so, do they have the resources they need?
- It is economically worthwhile to grow food in a certain area?
- How will a proposed venture contribute to urban sustainability initiatives?

Research into urban and non-commercial agriculture can address some, if not all, of these questions and can contribute to agricultural knowledge in ways that are currently overlooked.

Measure Benefits and Harm from Agricultural Systems

As important as it is to understand the benefits from an agricultural operation, the harm that its practices impose on society, the environment and the local ecology cannot be overlooked. By ignoring the negative impacts from a farming operation, the accumulated damage may eventually become greater than can be mitigated and the problems may affect other properties. Just as it is important to understand the magnitude of benefits from an agricultural operation, it is prudent to also know the magnitude of negative impacts it causes. By comparing the two, one can understand the *net value* provided by the operation. This also allows for a more robust comparison between different agricultural methods and operations, permitting the farmer to make more informed choices with respect to crop growing methods. The exercise of comparing *gross benefits* (all benefits, regardless of type) against *gross costs* is at the heart of measuring and understanding how sustainable any practice is or can be.

Some of the most pressing issues facing humanity today are pollution (air, land and water), depletion of natural resources (energy, water, raw materials), the demands of a rapidly increasing population and loss of natural habitat along with the environmental goods and services they provide. The United Nations Millennium Declaration (United Nations 2000) recognized the need to address global environmental issues and included a section entitled "Protecting Our Common Environment" that described the need to support the principles of sustainable development and resolved to commit to the adoption of an ethic of conservation and stewardship with respect to water resources, forest resources, biodiversity protection and greenhouse gas reductions. Because agriculture occupies a significant amount of the land area utilized by humans, it also plays a significant role in these issues. Understanding how the different forms and methods of agriculture can contribute to or relieve some of these pressing issues is a central reason why it is important to quantify both the negative impacts and the benefits of agricultural systems.

Not all agriculture is practiced by commercial enterprises. In fact, non-commercial interests, especially in the realm of human settlements, practice a good deal of agriculture. The products from those activities (home food gardens, community gardens, school gardens, etc.) usually do not enter the market system and the amount produced is often undocumented. The benefits from non-commercial agricultural operations are poorly measured and the contributions they make to a community's wellbeing are mostly understood qualitatively.

The American Community Garden Association (ACGA) website estimated that there are 18,000 community gardens in the United States (ACGA 2016). As the number of community gardens continues to rise, it is important to examine the practice to determine how it may be positively or negatively affecting the community and environment. Only then will it be possible to address negative impacts, suggest better management practices and implement more sustainable methods.

The world's population has grown over the past two hundred years, with the highest rates of increase occurring from 1950 to 1980 (Table 1). According to the US Census Bureau (2016), the world's population increased from 4.4 billion in 1980 to more than 6.8 billion in 2010. By 2050, Earth's population is projected to increase by an additional 2.5 billion. With this explosive rise in population comes an equally large increase in global food demand. At the same time, the extent of productive agricultural land is shrinking due to loss of arable land to urban development, desertification, soil loss, salinization, pollution, climate change and other factors.

Table 1. Estimated world population from 1800 - 2050.

Year	Population (billions of people)	Decadal Increase in Population (billions)	Average Annual Growth Rate (%)
1800*	1.0	-	-
1850*	1.3	-	-
1900*	1.7	-	-
1950**	2.56	-	1.75
1960**	3.04	0.48	2.00
1970**	3.71	0.67	1.8
1980**	4.44	0.73	1.74
1990**	5.28	0.84	1.42
2000**	6.08	0.80	1.20
2010**	6.86	0.78	1.07
2020**	7.63	0.77	0.87
2030**	8.32	0.69	0.67
2040**	8.90	0.58	0.52
2050**	9.37	0.47	-

*Data source: United Nations 2016
**Data source: US Census Bureau 2016

With the expansion of industrialized and high-intensity agriculture during the 20th century has come a long list of negative impacts (Nemecek *et al.* 2007); these include:

- Aquifer depletion
- Depletion of non-renewable natural resources
- Soil erosion
- The creation of toxic emissions that can end up in food (e.g., heavy metals, pesticides)
- Physical and biological impacts to the soil, such as soil compaction and organic matter depletion
- Changes to landscape structure and topography because of land alteration and intensive farming methods
- Changes to biodiversity
- Production of offensive odors
- Noise

In the past decades, there has been more awareness about the negative impacts arising from some kinds of farming operations. Monitoring programs that discovered how pollution in agricultural runoff caused downstream environmental problems caused this awareness. Once the connection between agricultural runoff and environmental degradation was understood and quantified, Best Management Practices (BMPs) were developed and implanted to reduce or eliminate impacts. These activities have made significant and important advances in protecting natural resources, and have become common business practices in some agricultural sectors. This is one example of how identifying and measuring harm from an agricultural operation can lead to successful solutions. Unfortunately, BMPs have been focused primarily on rural commercial agricultural operations. Urban and near-urban agricultural systems can also produce negative impacts. The need to identify, quantify and mitigate these will only grow as the practice itself continues to expand.

Meeting the world's food demand will be an enormous challenge, but can only be achieved by adopting sustainable food production methods, conserving resources and protecting natural ecosystems along with the services they provide to society. Humanity is becoming increasingly dependent on compact, efficient and productive agricultural systems that must also preserve natural lands and resources. It is not possible to judge one agricultural system as being a better option than another without analyses to document the inputs, outputs, benefits and losses associated with a particular practice. Understanding the full range of harm (often referred to as costs) and benefits arising from different agricultural systems is essential as the world's population continues to grow.

Inform Planning and Policy Decisions

Throughout the 20th century and early 21st century, the availability of locally grown food in the United States declined while population increased. Most of these new residents lived in cities, which caused the expansion of metropolitan areas and the development of urban sprawl (Szlanfucht 1999, Daniels 2009). Urban expansion often occurs at the expense of adjacent agricultural lands, because land that is suitable for farming is also desirable for development (Szlanfucht 1999, Lacy 2006). Concerns about the loss of farmland to urban sprawl have given rise to growth management policies and tools to preserve farmland in areas where development pressure poses the greatest threat (Lacy 2006, Daniels 2008). Although some farmland preservation tools have been successful, most regional and state comprehensive planning efforts have been generally ineffective (Szlanfucht 1999, Lacy 2006, Daniels 2008). Studies on the potential benefits of urban agriculture are needed to better guide urban planning initiatives.

Urban developments are typically planned with considerations of safety, aesthetics, community activities and access to mass transit, but urban planning has traditionally excluded food-producing areas. Developers often view food-producing lands within urbanized areas as open land that is ripe for the establishment of residential or commercial projects. The assumption is that urban residents are expected to be fully dependent on an imported food supply, relegating the function of food production to rural farms. The ruralization of food production has increased the distance that urban residents' food has traveled, carrying with it a high energy and carbon dioxide (CO_2) footprint. Urban planners may be more receptive to urban agriculture within their planned communities if they understand how it can positively contribute to the community.

Urban planners and community leaders who wish to increase community sustainability through a wider practice of urban agriculture are often faced with the need to demonstrate the potential benefits of proposed initiatives. Government officials often need to quantify the benefits of a sponsored activity or to demonstrate the return on the investment in an urban agriculture program. Quantifying the benefits, in both type and magnitude, allows officials to make decisions about how urban agriculture can fit into urban planning goals and objectives.

Urban agriculture can be more easily integrated into the planning process if knowledge of its benefits, feasibility, forms and receptivity by residents are known. A serious and scientific study of urban agriculture can provide such insight and will help bring an informed discussion about urban agriculture to the planning process.

Support Sustainability Initiatives

Agriculture is directly connected to many issues that are the focus of sustainability initiatives. These include:

- Non-renewable energy consumption
- Greenhouse gas emissions to the atmosphere
- Environmental impacts from farm runoff (e.g., pollution of waterways and groundwater)
- Depletion of non-renewable resources
- Decline in the extent and quality of natural habitat, which contributed to species extinction
- Unsustainable farming practices that lead to soil depletion and degradation
- Food insecurity and food deserts
- Better use of underutilized urban land

Why Quantify?

Many cities and regional governing bodies have created sustainability initiatives that are aimed at conserving resources, reducing operating costs and reducing the community's ecological footprint. For most, the rationale behind these initiatives is that resource conservation (i.e., water, non-renewable energy, etc.) will have long-term benefits by reducing the maintenance and operating costs. In regions where it is expensive to develop new potable water sources, water conservation is an effective way to avoid the costs of expanding municipal water infrastructure. Building a new power plant comes with high costs (which are passed on to consumers) and a lengthy permitting process, both of which can be avoided if energy conservation measures hold the line on additional demand. More often than not, sustainable and local agriculture has not been seen as something that is connected to these and other resource conservation initiatives although it indirectly is.

Sustainability initiatives across the United States include a varied list of programs. Some of these are shown in Table 2. Urban agriculture is associated with many of these elements, often in unexpected ways. For example, food gardens at educational institutions can be integrated into a wide range of course topics, including mathematics, biology, ecology, economics and environmental studies. Urban agriculture can play an important role in creating wealth, which benefits the local community and supports economic health. Urban food gardens, when appropriately designed, positively contribute to the urban ecology. Native wild food plants, which are a part of almost all natural plant communities, can be a valuable feature of urban nature parks and preserves. Some urban food gardens are very efficient water users and are a water-saving option over irrigated greenspace, thus providing a measure of water conservation. Local food production reduces the amount of food that needs to be imported into the urban setting, which conserves energy and reduces greenhouse gas emissions associated with transportation. Lastly, urban agriculture can recycle a large volume of waste materials that would otherwise be placed in landfills (e.g., building materials) and can recycle the organic matter and nutrients found in compostable waste generated by the food service industry.

Table 2. Elements typically included in sustainability initiatives.

Programs	Targets
Education	STEM expansion in colleges/universities, Smart grid & smart city research programs, cultural events and activities to promote sustainability awareness
Economic Development	Clean technology investment
Urban Greenspace	Develop new greenspaces,
Urban Natural Habitat	Tree planting programs, waterway/water body restoration
Wastewater Reuse	Water resource conservation
Stormwater	Green storm water infrastructure investment: permeable surfaces, rain gardens and above- and below-ground water storage
Water Conservation	Conservation initiatives and education, upgrade/replace outdated water mains
Energy Efficiency	Retrofit buildings, solar energy investment, fuel-saving vehicles, investment in mass transit
Clean Energy & Greenhouse Gas Emission Reduction	Investment in clean energy sources
Recycling	Expansion of recycling programs, reuse retired infrastructure materials
Transit & Accessibility	Bike Sharing Program, pedestrian-friendly streets, rideshare program
Local Food System	Expansion of local and fresh food offerings through markets; develop linkages between food assistance programs and local farmers markets, locally-sourced food in schools
Urban Food Production	Support for community gardens, rooftop gardens

Chapter 3. An Overview of Agriculture

Humans obtain food through two primary channels. For one, food can be gathered from natural ecosystems. From the earliest period in human history, this was the sole source of food for ancient cultures and still is to many indigenous peoples who live within their ancestral lands. Food is also obtained by the cultivation of other living things, which is referred to as agriculture. At its core, agriculture is the practice of growing organisms for their products, which are used to support human health, nutrition and needs. Agriculture involves the cultivation of a multitude of organisms for a wide range of products. Some of these organisms and agricultural products are listed in Table 3.

Table 3. Examples of agricultural products.

Type of Agriculture	Organism	Products	Examples
Animal agriculture	Cattle, fowl, fish, etc.	Human food, byproducts for animal food or fertilizer	Beef Manure
Fungiculture	Fungi	Human food, medicines	Mushrooms Penicillin
Microbial agriculture	Bacteria	Pharmaceutical and natural pest control products	*Bacillus thuringiensis*
Agronomy	Plants	Food (human and non-human), ornamentals, fiber, fuel, land reclamation, environmental restoration	Beans Flowers Cotton

The science and practice of agriculture cuts across many fields of study; some of these are:

- Biology, especially the areas of plant biology and plant physiology
- Genetics, particularly related to improved crop yields and disease resistance
- Chemistry, including biochemistry and inorganic chemistry
- Ecology, at the local and landscape levels
- Earth science, including meteorology, geology, soil science and hydrology
- Environmental planning
- Economics
- Sociology, especially with respect to urban agricultural systems

- Food systems

Agriculture is a broad term that includes different applications. For this reason, it has been divided into sub-disciplines that define areas of specialization, such as plant breeding, research and development of agricultural chemicals for fertilizers and pest control (agrochemicals), development of genetically modified organisms (agricultural biotechnology), production techniques, ecological approaches to agriculture (agroecology), and the science and technology of producing and using plants (agronomy). Within each of the sub-disciplines, there are specialized areas of study and much overlap with other fields of study. For example, areas of specialization in agronomy include pest management (weeds, pest insect control and disease prevention), productivity optimization, crop rotation, water management (irrigation and drainage), crop plant studies (plant physiology and breeding), soils (classification, fertility and erosion control) and more recently, BMPs.

The primary focus of this book is on the practice of growing food plants, particularly as it is practiced within cities for the purpose of food production. One may argue that this book is also concerned with the topic of agroecology because it has a heavy focus on sustainable agricultural systems that are compatible with sustainable ecological systems. The other sub-disciplines of agriculture are excluded, not because they are less important, but because this narrowed focus allows for a more manageable scientific study.

The act of cultivating plants requires one to understand what can reasonably be grown under a given set of conditions. Agricultural yields can vary greatly between different sites and there is a strong seasonality of agricultural production in most regions (Nemecek *et al.* 2007). Growing conditions change from year-to-year and agricultural production is sensitive to these types of uncertainties. Although some uncertainties can be managed, others cannot; successful agricultural production requires understanding the variables and how these can be managed to maximize production while meeting objectives and goals. Some of the uncertainties that influence agricultural production include, but are not limited to:

- Availability and reliability of labor
- Prices of required materials
- Occurrence of catastrophic events
- Soil conditions
- Water availability
- Weather conditions
- Appearance of weeds, insect pests, pathogens and diseases

The success of any agricultural operation requires that enough products can be harvested (or benefits realized) to make it worthwhile. In addition, what constitutes the definition of "worthwhile" is not always straightforward. This last point suggests that the benefits that have been received from the agricultural operation equals or exceeds the costs of the activity. One grower may value the tangible products from his or her labor (e.g., food production, creation of an asset), but for other growers there are benefits that are not so readily recognized- particularly those that are defined as intangible benefits. There is more discussion about intangible benefits in later chapters.

There are significant costs (i.e., inputs and consequences) associated with the practice of food growing. As with the benefits, the list of costs includes tangible things like financial investment and pollution caused by the growing operation. There are also intangible costs. Over the past decades, a number of environmental costs (impacts) associated with agriculture have been studied in-depth. Although the weighing of benefits against costs is a core concern for practitioners of agriculture, there are other issues that are equally important but are often not considered. These will be discussed in later sections.

The list of benefits, costs, harms and uncertainties vary by agricultural type and the growing methods used. Agricultural production methods have changed profoundly over the past century and there are many more different methods that have been developed and implemented. Below is a summary of the different kinds of agriculture and how they contrast with each other, with a particular emphasis on fruit and vegetable production.

Commercial and Non-Commercial Agriculture

There are profound differences between commercial and non-commercial agriculture. The home food gardener, community gardener, small family-owned subsistence farming operation on the urban fringe and for-profit agri-business all see agriculture as the practice of growing food. But the factors of *how* that food is grown, *why* it is grown, the *consumers* of the agricultural products, the *economics*, the relationship to the community, the *tangible benefits* and the *intangible benefits* of commercial and non-commercial agriculture are very different. These differences are why the extent of commercial agriculture is far in excess of non-commercial agriculture, governmental subsidies are focused on commercial agriculture and profitability is variably defined by the influence of each factor.

Commercial

Commercial agriculture exists to produce food for commercial markets and these markets exist to produce food for those who do not grow it for themselves. In essence, commercial agricultural businesses provide a service for a fee. Because food is being grown for profitable sale, the crops being cultivated are generally those that are likely to yield a profit for the grower.

Until the 1900s, market farms on the lands adjacent to cities were the primary commercial food suppliers to most urban residents. Most food was grown locally and consumed within the region where it was produced, and was usually sold by the farmer (Roberts 1992). With the advent of industrialized agriculture during the late 1800s, commercial agriculture began to change. Between 1900 and 1930, commercial agricultural production in the United States shifted to western states because of irrigation subsidies (Roberts 1992). With the advent of national railway and highway systems, refrigerated freight cars, chemical pesticides and industrial machinery, truck farming emerged as the major farm type beginning in the 1950s. Truck farms typically grew few types of crop, were located long distances from the point of product consumption and relied on non-renewable energy sources for production.

After the 1950s, agricultural production in the United States changed dramatically with commodity agriculture and regional specialization becoming established practices (Lyson & Guptill 2004). Although consumers benefitted from year-round supplies of fresh fruits and vegetables in stores, they were also becoming more distanced from agricultural production. During this time, the practice of urban agriculture waned as planning policy focused on rural protection, urban containment and agricultural policies that stimulated rural food production (Marsden *et al.* 1993). Some of the changes that took place in the United States, Canada and Great Britain included the intensification of productivity, specialization of produce, specialization of labor and the concentration of farm and land ownership away from family farms (Ilbery, Chiotti & Rickard 1997).

With the emergence of large agribusinesses, agricultural policies shifted in favor of large agribusinesses and the overproduction of commodity crops at the expense of the environment and small local farmers (Peters 2010). The current commercial agricultural system is heavily supported and subsidized by the United States Department of Agriculture, land grant universities and large multinational agribusinesses (Lyson & Guptill 2004). During the early part of the 21st Century, an estimated 98 percent of the United States food supply came from industrial agribusinesses that used intensive farming practices (Peters 2010).

Commercial urban agriculture can take many forms that include small farms and urban food production facilities. Urban (and near-urban) farms usually produce food on smaller plots of land, sell to (mostly) local customers and use intensive farming methods. Urban food production facilities, which sometimes advertise themselves as urban farms even though they do not look anything like a typical farm, are often found in greenhouses, renovated buildings and warehouses in larger cities. Because food is being grown indoors, the facility can resemble a factory with various production lines that are monitored and controlled by technology. Urban food production facilities often employ industrialized methods, use intensified cropping, grow using hydroponics, rely on artificial lighting and specialize in fast-growing crops. Some of the most common crops produced by these facilities are tomatoes, sprouts and different varieties of lettuce.

Non-Commercial

Non-commercial agriculture includes a variety of food growing practices that are all connected by a single theme- the food that is produced is not intended for sale and, therefore, is not subject to or is only minimally concerned with market influences. In the absence of market and profitability constraints, there is a greater diversity of crops that are grown for a surprisingly broad range of purposes. In contrast to commercial agriculture, which is focused on a single or very few crop plants, non-commercial agriculture can be more reflective of the grower's needs and wants.

Examples of non-commercial agriculture include:
- Home food gardens
- Community gardens
- Guerilla gardens
- Subsistence gardens or farms
- School or educational gardens

Subsistence farming and home food gardens are typically operated to provide food and other plant products to the family that works the garden plot. Community gardens and guerilla gardens are often focused on producing food for local residents. The crop production methods used are often, but not exclusively, related to the size of the land that is being cultivated.

Production Methods

There are many books, articles, websites and other resources that describe the many different types of agriculture being practiced today. Urban agriculture encompasses a wide variety of production methods. In order to understand how these are related, the history of their development and the general characteristics of each, a review of the different types of agriculture is warranted. Be aware that many authors have presented different schemes to classify the different types of agriculture and the methods used to grow food. The one presented below is not presumed to be superior to those presented by other authors, but it is a classification scheme that best serves the objectives of this book.

Before going further, it is important to clarify some nomenclature with respect to the use of the terms "agricultural types" and "growing methods". For the purpose of this book, an agricultural type will refer to the grounding principles that guide the food growing operation, which includes its goals and objectives. A growing method is the specific activity that is used to produce the food. To illustrate the relationship between the agriculture type and growing method, commercial agriculture (a type of agriculture) is practiced to generate profit but can use any number of food growing methods. In another example, organic agriculture is an approach to food production that uses only natural fertilizers and pesticides. The actual growing methods used may vary from region to region, but they all adhere to the grounding principles.

Three main types of agriculture (i.e., those most practiced) are:

- Traditional
- Organic
- Industrial, including:
  - In-ground (field)
  - Hydroponic
  - Container

Traditional

Traditional agriculture is as old as the practice of agriculture itself. The term typically refers to food growing methods that were used before the advent of industrial agriculture and is sometimes associated with the terms "subsistence farming" or "subsistence agriculture." Traditional agricultural practices are often small in scale, locally adapted, non-technological, ecology based and have characteristics that support ecological and social sustainability (González Jácome 2009). The sustainable nature of these operations is mostly based on the fact that they are small scale and they often occur in areas where there is a shortage of agricultural land- characteristics that make these practices especially appropriate for urban settings, which have similar limitations and needs.

Traditional agricultural methods were widely practiced in most non-nomadic cultures until the 20th century, when industrialized growing methods were being developed and promoted by governments in developed or developing countries as a means to feed growing populations. By the end of the 20th century, concerns about the ecological impacts of large-scale agriculture on the natural environment spawned a greater interest in the science of agroecology. Agroecology initially focused on the ecological processes that operate within an agricultural system, but the discipline has expanded to include considerations of how an agricultural system is integrated with the local ecology. This shift in focus recognized how some of the ancient agricultural practices of indigenous peoples also protected biodiversity while providing for a sustainable harvest of products from natural ecosystems.

Some examples of traditional agricultural systems include:

- Amish, in North America
- Balinese, in Indonesia
- Kogi, in Colombia
- Inca, in Peru

The Amish are a religious group that holds simplicity as central to their core beliefs. The most conservative members shun ornamentation, technology and self-importance. What is most striking is that their way of living, and particularly their way of farming, has remained unchanged over hundreds of years, essentially ignoring the Industrial Revolution. Cornerstone to their agricultural practices is the use of horse or oxen-drawn plows, natural manure and compost fertilizers, and natural (organic) methods of vegetable production. The Amish typically live on small subsistence farms and supply most of their own basic needs. They recycle and reuse whatever is available to them and have adapted their lifestyle to live within the environmental means available to them without importing energy or resources from other regions. (Orr 2002)

The traditional farming practices of the Balinese developed over thousands of years and are sustained by a ritual-based system of irrigation, pest management and planting schedules. Water temples administered by irrigation cooperatives manage water resources. Farmers meet at temples to discuss the condition of their fields and pest situations. They coordinate planting schedules according to the indigenous calendar, which regulates the irrigation schedule for numerous farms in a watershed. This system ensures benefits to all farmers by synchronizing water allotment to manage pests. This system of synchronized planting has proven its efficacy over modern agro-industrial methods based on pesticides and fertilizers. Farming practices form an intricate web of religious, social and technological relationships. Interestingly, in the 1960s, the water temple system was abandoned when new varieties of rice and growing practices were adopted. At first there was an increase in yield but that soon declined, the water began to become scarcer and pests became a problem. Fertilizers degraded water quality and polluted the environment. In the 1980s, the Balinese began to return to their traditional farming system (Edwards 2010).

The Kogi in the South America mountains of the Sierra Nevada de Santo Marta (Colombia) and the Inca in the Peruvian Andes share many similarities in their agricultural systems. Both live in mountainous regions with rugged and steep terrain. The Inca created extensive networks of vertical terraces that were part of an interconnected irrigation system, which was fed by natural streams and springs on high slopes. Because of the landscape's verticality, both the Kogi and Inca planted crops along the elevation gradient along a mountain's inclines. This permitted the planting of a high diversity of crops at various elevations- manioc, corn, sugarcane and pineapple in the lowlands, and potatoes and onions in the midlevels, where they also graze domesticated animals. Fertile sediments from floodplains at the base of the mountains were carried up to the agricultural terraces and food growing plots, providing a natural source of fertilizer. These agricultural systems are compatible with the local environment because they incorporate the growing site's natural characteristics to cultivate a variety of native crop plants, utilize natural and renewable sources of plant nutrients, and do not deplete the resources they depend upon (such as water).

Organic

Organic agriculture, as a formal agricultural concept, was developed during the 20th century as a revival of the ancient agricultural practices that were used to improve and maintain soil quality (e.g., crop rotation, growing cover crops), fertility (applying animal manure, green manure, compost and fertilizers derived from plant or animal sources) and pest management (biological pest control, disease control and prevention, and application of pesticides from natural sources). Fertilizers and pesticides derived from mineral sources, laboratory processes or manufacturing activities are avoided. Besides these restrictions, organic agriculture is also practiced out of a respect for sustainability, resource conservation, environmental protection, ecological integrity, health and food safety.

One of the goals of organic farming is to increase total system biodiversity (Pimentel *et al.* 2005), which can be achieved by planting a range of crops, using crop rotation, prudent use of pesticides and reserving natural areas. Significantly greater abundances and species diversity have been found on farms that use organic rather than conventional growing methods (Hole *et al.* 2005, Fuller *et al.* 2005).

The tie between traditional and organic agriculture is very close, but has become more separated over the second half of the 20th century as new growing techniques have been developed that are not "traditional" but are "organic". Like organic agriculture, traditional agriculture relies on ancient methods for food production; however, some specialized commercial and non-commercial agricultural operations (that are nothing like traditional agriculture) also grow according to organic production guidelines. Most commercial agriculture does not grow using organic practices (organic products account for less than 5 percent of the total food sales in the United States). It is likely that a higher percentage of non-commercial agriculture uses organic practices but reliable survey data to investigate this is lacking. One reason that organic food production is not a major part of the commercial food system is because it is viewed as being less productive and less profitable than more intensive farming practices. However, long-term studies that compare side-by-side crop production from organic and conventional high-intensity farming methods have shown that this is not necessarily true (Hepperly, Douds & Seidel 2006, Pimentel *et al.* 2005, Reganold, Elliott & Unger 1987).

One can find several variations for the definitions of "organic agriculture". Non-commercial agriculture practitioners, such as home and community food gardeners, accept looser definitions. In the stricter sense, legal definitions have been established by governing authorities and these define specific guidelines that must be met in order to sell produce that is "certified organic" in the commercial market system. More recently a new label has emerged that identifies food as having been produced using organic methods or organic principles, rather than being certified organic. This allows the producer to sidestep the sometimes rigorous and often costly process of being certified as an organic food producer, while notifying the consumer that the grower adheres to some type of organic production ethic.

Industrial

Industrialized agricultural production is also referred to as intensive agriculture and by the latter part of the 20th century it became known as "conventional agriculture." Conventional agriculture is the basis of the modern commercial food market system. The growth of agricultural productivity during the 20th century is attributable to the mechanization of farming activities and intensive use of fertilizers and pesticides (Nemecek *et al.* 2007). During the early part of the 21st Century, an estimated 98 percent of the United States food supply came from industrial agribusinesses that used intensive farming practices (Peters 2010). Industrial agriculture relies heavily on crop production techniques that are designed to maximize output; these include heavy use of chemical amendments (plant growth regulators, fertilizers and pesticides), focus on one or few crop types, growing multiple crops per season and the use of mechanized planting/harvesting equipment. Some researchers consider any food production method that involves a controlled growing environment and relies on manufactured materials and modern technology to be a form of industrial agriculture. Under this definition, food growing using greenhouses, hydroponic systems and containers would fall under the umbrella of industrial agriculture.

Industrialized growing methods have been associated with unsustainable practices, including a high consumption of natural resources, high use of pesticides, monoculture crops, dependency on non-renewable energy sources, soil degradation, greenhouse gas emissions to the atmosphere, unsustainable use of water resources and pollution from runoff. Although some of these impacts can be mitigated or reduced, the extent of natural habitat that has been lost to large-scale agricultural operations is enormous and cannot be offset. The negative consequences of industrialized farming also extend to many aspects of society, including rural life, environmental quality, public health, governmental regulations and markets (Peterson 2000).

Hydroponic

Hydroponics, a type of container gardening, is the practice of growing (crop) plants in a soilless medium (e.g., perlite, gravel or manufactured material) or hydrologic solution. The fertilizers used to support growth are usually from mineral (chemical) sources, although organic hydroponic solutions have more recently become available to growers. Hydroponic growing methods usually offer a disease- and pest-free environment to grow in, potentially offering much higher yields and more sanitary conditions than could be created in fields or garden beds. Hydroponics are a desirable alternative for those who live in areas where the climate is too extreme to allow traditional outdoor food production or where there are insufficient soil resources.

Some forms of aquaculture, such as fish production, are integrated with hydroponic food crop production. Integrated fish-crop production systems produce both animal protein and food crop products within the same production area. For example, rice and crayfish can be produced from the same impoundment or large container. There are efforts to develop self-contained systems that include fish production tanks whose water is recycled for food plant production with minimal outside inputs.

Container

Food is grown in containers, rather than in a garden bed or farm field. Container gardening can be used for both commercial and non-commercial food production. These can be fixed (vertical gardens, rooftop gardens) or mobile (patio gardens). Greenhouse production is usually done in containers. One particularly inventive container garden can be found in the city of Edinburg, Scotland where each food-growing container sits on a recycled wood pallet. These containers can be moved around with a forklift or loaded onto a truck for transport to a new location when necessary. The community garden occupies a site temporarily and can move to a new space once access to a site is lost, allowing both the garden and the landowner flexibility with respect to commitment.

Crops

A seemingly endless variety of crop plants are available for food production. Beyond the seed catalogs, the number of naturally occurring (wild) food plants available for cultivation is staggering. Unfortunately, many of the world's best traditional food plants have been forgotten. In the horticultural world, the types of crop plants that are available to growers can be placed into several broad categories, based on what the crops are used for or the type of agriculture that grows them. These are described below.

Traditional crops

The term "traditional crops" does not refer to the crops produced by traditional agriculture; traditional crops are those that are typically grown by commercial agricultural operations. Many of these crop types were developed and have risen to prominence during the 20th century. Field crops are those that are intensively grown on larger tracts of land, such as corn, soybeans, wheat, oats and sorghum. These are usually planted in close rows early in the growing season and harvested after the plants have seeded and dried. In some places, conditions allow for multiple crops to be grown in a single season. To maximize yields, these crops are produced using high-density growing methods that include the heavy use of chemical amendments (plant growth regulators, fertilizers and pesticides), focus on one or few crop types and the use of mechanized planting/harvesting equipment. A much smaller number of traditional crops are produced using traditional agricultural production practices, usually in the milieu of subsistence farming.

Specialty Crops

Besides traditional crop types, there are other food plants (and plant products) that are grown for commercial markets and non-commercial use. These include herbs, flowers, fiber plants, fruit, vegetables and berries. Usually, specialty crops are grown in relatively smaller quantities than traditional crops because they may have a lower demand or they are more labor intensive to produce. Specialty crop production is sometimes restricted to specific climatic regions or demand is associated with certain cultural groups and the crop is predominately sold in niche markets.

Subsistence Crops

Subsistence crops are those that are grown to meet the food needs of those who depend on the operation. Specific varieties are selected to include a range of vegetable types, such as high-energy vegetables (starchy), greens and others that support a healthy diet. Two unique aspects of subsistence crops are that they vary from farm to farm and they are grown for reasons other than commercial production (i.e., profit making). Cultural food preferences usually play a large role in determining the primary vegetable crops that are grown (Carmichael 2011). In temperate climates, where there is a strong seasonality to crop production, the types of crops produced may also be influenced by the methods available to preserve food for consumption during off seasons.

Crop & Livestock

Although livestock and poultry production is beyond the scope of this book, some crop plants are specifically grown for livestock feed or used both for human food or animal food. For example, field grasses and other green plant material are grown and harvested for silage. Some cover crops are grown for browse material. These animal crop plants can be produced using traditional, organic or industrial practices and may be associated with commercial operations. Sometimes livestock feed crop plants have been selected and seeded to create a specific diet; in other cases, fields are left fallow and the resulting growth is a reflection of the soil seed bank.

Unusual Crops

Another type of small-market or low-demand crop type is the unusual crops, which are often small animals that are used for specific purposes. These are animals, but these have a more utilitarian purpose and are used to benefit plant crop production in some way. Some unusual crops are:

- Butterflies, for butterfly gardeners (crop plant pollinators)
- Bees, for establishing new hives (crop plant pollinators)
- Agriculturally-beneficial insects (e.g., ladybugs, praying mantis), for sale to organic agriculture operations
- Worms for composting

Scale and Settings

Agriculture is practiced in a wide range of settings, from farm fields to condo patios, from community gardens to residential backyards. Each of these different production sites has unique benefits, challenges and environmental conditions that affect the quantity and quality of the food that is being grown. Below are summaries of the different settings in which urban agriculture is most often practiced.

Large-Scale Commercial Farms

Large-scale farms are those that grow crops for commercial markets and require fuel-powered machinery to manage. Because of cheaper land prices and the lack of larger land parcels within or adjacent to metropolitan areas, these farms are typically located within rural areas. This distance from the consumer requires harvested produce to be transported to urban markets, which adds cost to the farm product that can offset the lower operational cost. Large-scale farms are often the least productive per unit area and carry a large ecological footprint due to the resources they consume and their environmental impacts.

Small-Scale Farms

Small-scale farms are those that may be managed by small machinery or work animals. These farms may be subsistence farms, market farms or part of a co-op that produce for a larger market. Small-scale farming is generally more productive and efficient than large-scale agriculture, indicating a greater potential output per unit area than would be achieved through conventional agricultural methods. Carter (1984) found that the per-hectare farm production in India declined 20 percent as farm size doubled. Studies indicate that the reasons for this drop in efficiency with increasing farm size may vary by region (Carter 1984, Barrett 1996, Assunção & Ghatak 2003, Lyson & Guptill 2004). Some authors suggested that this higher efficiency of small-scale agriculture may be attributed to the lower costs of supervision, decreased land to labor ratio and better use of resources (Assunção & Ghatak 2003, Ghosh, 2010).

Community Gardens

Communal or community gardens usually occupy a piece of land that has been made available to local residents for the purpose of food production. Because community gardens exist for a number of reasons, they also come in many different sizes and styles; some are sites of community events. Depending on the landowner and bylaws of operation, these gardens may have any of the following restrictions:
- Food is produced for the consumption of the garden's participants only
- Harvested food, or a portion thereof, is donated to local charitable organizations
- The garden is run for education, research and demonstration purposes
- Produce may not be commercially sold

Home Food Gardens

Residential fruit and vegetable gardens are established to grow food for the household and are highly influenced by the background, preferences and heritage of the gardener. These vary widely in size, practice, purpose and what is being grown.

Some home food gardens may be nothing more than pots of herbs or a few vegetables grown in containers on a patio. Others are more established fixtures on the property and may include raised beds, a utility area (tool storage, compost pile, etc.) and an irrigation system. Larger gardens may contribute significantly to the household's annual food budget by providing most of the fruits and vegetables that are needed. Some home food gardens are also important areas for family gatherings and other social activities.

Chapter 4. Sustainable Urban Agriculture

Over the past decades, there has been a fundamental shift in agricultural research focus and food production methods. During the 20th century, agricultural science was generally preoccupied with maximizing productivity (Peoples, Herridge and Ladha 1995). This was accomplished by advancements on several fronts: (1) plant genome manipulation, (2) plant nutrition (3) pest control and (4) expansion in the ways and places that food plants can be cultivated. Much of the drive to increase agricultural productivity came from the need for commercial operations to be economically viable.

By the later part of the 20th century, it became increasingly clear that a narrow focus on food plant cultivation, harvesting and efficiency had excluded critically important factors that are also part of agricultural systems. Some of the negative impacts of large-scale rural farming operations, which are the backbone of the commercial food system, started to be identified and measured. These included pollution runoff, environmental degradation, unsustainable use of resources, habitat loss, loss of knowledge about how food was grown and people's disconnection from the land. As these problems became known, scientists began to explore the concept of sustainable agriculture and define its principles.

What is "sustainable urban agriculture," what are its characteristics and why is it important to warrant serious scientific study? To answer these questions, it is best to deconstruct the term into its constituent parts to fully consider the inherent meanings of each. The word "agriculture" includes the very different practices of community gardening, commercial crop production and home food growing. These practices are all types of agriculture because they involve the cultivation of organisms for food or products. The topic of agriculture was the main focus of the previous chapter and will not be repeated here. The word "urban" indicates the specific place where the agricultural practice takes place and infers a unique set of conditions that are distinctive to the built environment. Lastly, the word "sustainable" is perhaps the least understood of the three parts of the term "sustainable urban agriculture." But in its simplest form, a sustainable practice is an activity that can be carried out indefinitely without damaging or depleting the resources it depends upon. The terms *sustainable* and *urban*, as they relate to agriculture, will be further discussed below.

Besides defining what sustainable urban agriculture is, it is also valuable to define what it is not. There are many agricultural practices (both urban and rural) that tout environmental, ecological or economic benefits, but one cannot ascertain how truly sustainable these are without an objective set of criteria to measure them against. Below is a set of criteria that will be used in this book to define the characteristics of a sustainable urban agricultural system.

Urban

Urban food production is carried out in an environment that is radically different from rural food production, where most of the commercial food supply is grown and agricultural studies have taken place. The physical characteristics of the urban setting and the strong influence of human interactions there have created conditions that are like no other. Some urban sites are contaminated with pollution, a legacy from previous land uses or deposition from nearby mobile sources. In addition, vacant urban land is usually under considerably more development pressure and carries a much larger tax burden than rural land- characteristics that make profitable urban farms a difficult venture. Cities can encourage urban agriculture on idle or underutilized land; however, recent trends towards more compact urban development (Gordon & Richardson 1997) reduce the inventory of potential production sites.

Urban agriculture can take many forms and can be practiced on public or private land. Different types of urban agriculture include public gardens, private gardens, edible landscaping, orchards, green roofs, aquaculture, small-scale farming and hobby beekeeping (Mendes *et al.* 2008). Perhaps the most visible example of urban agriculture is the community garden (Peters 2010), which consists of publically or privately owned land on which individuals are granted permission to access plots for food production. Community food gardens provide urban residents with locally grown fresh vegetables, social interaction and a stimulus for improving their community (Shutkin 2000, Corrigan 2011). These benefits are particularly important in low-income and underserved neighborhoods where access to fresh, healthy food is typically limited.

The dominant model of food flows and systems for cities includes a heavy reliance on food grown in rural areas and transported into urban retail markets. Several examples from the literature suggest that this is not necessarily a requirement to meet urban food needs. Urban agriculture, in general, and community and home food gardens in particular, can significantly contribute to local food production. For example, during World War II governments in the United States and Great Britain encouraged civilians to plant food gardens as a way to help the war efforts. These "Victory Gardens" covered approximately 300,000 acres and met half of England's fruit and vegetable needs during and after World War II (Garnett 1996, Martin &

Marsden 1999). In the United States, most major cities (e.g., Boston, New York, Chicago) established Victory Gardens on public lands and, combined with private backyard vegetable gardens, the number of food gardens reached approximately 34 million during that time (Alexander *et al.* 1999). Victory Gardens supplied an estimated 40 percent of the United States' fresh vegetables in 1944 (Bissett 1976). Unfortunately, the number of urban food gardens declined after WWII because of postwar affluence (Broadway 2009) and the expansion of industrial agriculture (Roberts 1992).

Broadway (2009) described the first North American postwar resurgence of civic agriculture in Montreal, Canada in the mid-1970s when the city established a community garden program. This grew to include more than 76 garden sites with 6,400 plot allotments and 10,000 participants. At the same time Seattle, Washington began its community garden program and by 2009 it had more than 70 organic gardens that produced between 7 to 10 tons of produce annually. In 1995, California began its school garden program and by 2009, there were over 3,000 school food gardens across the state. The City of Philadelphia's urban food gardens produced approximately 907,000 kg (2 million lbs.) of produce in 2008, which had an estimated value of $4.9 million dollars (Greenbiz 2015). The City of Detroit established the Garden Resource Collaborative in 2003 to provide technical information, marketing resources and seeds to urban food gardeners. By 2008, the program supported 169 community gardens, 40 school gardens and 359 family gardens. The estimated combined production from Detroit's urban gardens was over 163 tons in 2008. It was estimated that by 2014, Detroit had 1,400 urban food gardens and farms (Daily Detroit 2015) that were worked by approximately 20,000 people (Cranes Detroit Business 2016) and produced nearly 181,000 kg (400,000 lbs.) of produce (Greenbiz 2015).

The 20th century trend toward commercialism of food has led to a greater concentration and globalization of food production and distribution, which placed enormous power and resources in the hands of few large, multinational corporations (Roberts 1992). These commodities are sold through retail sales outlets, which restrict the consumer's access through their ability to pay and the location of retail stores. Although consumers benefit from competitive pricing, consistency in quality, the creation of jobs, local economic support through taxes and local real-estate investment, the products offered for sale are determined by market supply, distribution chains and profitability, causing the consumer to lose local control over food selection. The commoditization of food has contributed to the development of food deserts and food insecurity in cities as the availability of food has been reduced to specific outlets (Pothukuchi 2004).

The community food system, which includes urban and near-urban farms, community gardens and backyard gardens, is a more economically efficient and sustainable alternative to the current market-oriented food market system available to city residents (Sonntag 2008). Food production in urban settings can eliminating the energy consumption and transportation costs associated with large agri-businesses (Faist, Kytzia & Baccini 2001). Local and urban agriculture help cities reduce their dependency on externally produced products and can act as a catalyst for improving the urban environment. In England and Wales, the practice of community gardening has been considered as an important part of urban renewal (Martin & Marsden 1999). In Cuba, severe economic conditions have been the driving force behind the nation-wide implementation of home and state-sponsored organic food gardening (Warwick 1999). Today, nearly all of Havana's 2.1 million residents are fed from fruits and vegetables grown within a 30-mile radius of the city and Cuba produces its food using organic agricultural methods (Altieri *et al.* 1999, Bahnson 2010, Peters 2010). The United Nations' Food and Agriculture Organization has reported that worldwide, some 800 million people grow food in cities and produce up to 15-20% of the world's food supply (Greenbiz 2015).

The Physical Environment

Although less visible than community gardens, private greenspace (usually lawn associated with residential areas, parks and institutions) is one of the largest land cover types in cities (e.g., Randall, Churchill & Baetz 2003, Gaston *et al.* 2005). These spaces offer the greatest potential area for urban food production. Residential spaces that can be suitable for agriculture include backyards, rooftops, patios and balconies (Smit & Nasr 1992). Private greenspace lends itself to urban food production because it does not have some of the problems of large-scale or commercial agriculture (e.g., labor costs, land costs, profit margin, taxes, and market fluctuations) and remains secure in terms of access and rights. Residential home food gardens may represent the greatest potential area for local food production in most metropolitan areas because of the development pressure on urban, suburban and near-urban farms as well as the limited availability of urban land for agriculture (Aubrey *et al.* 2012).

The urban environment is unique and can vary from parcel to parcel. The distribution and extent of greenspace, paved surfaces and buildings is relatively unequal from place to place and sometimes exhibit random patterns. Each of these components influence the site-specific environment, creating a mosaic of conditions that may variously favor one type of crop or hinder the growth of another. Because of this, there is a more diverse set of growing conditions present within cities and crop production recommendations that have been developed from field studies may not be reliably applied to the urban setting.

One characteristic of many densely urbanized landscapes is the condition of "extreme heterogeneity," which is a high degree of variability over a relatively small spatial scale. Extreme heterogeneity creates a perplexing matrix of growing conditions (e.g., exposure, temperatures, light intensity), minimizing the usefulness of growing recommendations that are usually available from agricultural information sources. This heterogeneity can also have important implications for the urban ecology, which is intimately tied to urban agricultural production through the sharing of agricultural pests (e.g., weeds, plant diseases, insect pests) and beneficial organisms (e.g., pollinators, pest predators).

The collective area of urban paved surfaces, particularly those that are darker tones, absorbs and stores heat during the daytime hours. This can raise daytime temperatures, particularly when there is little or no wind, creating a condition that is referred to as the "urban heat island" effect. During overnight hours, thermal energy that was absorbed during the daytime is released (radiated), causing local air temperatures to be warmer than they normally would be. According to the United States Department of Environmental Protection (2016), the annual mean temperature for a city of 1 million people or more can be 1° to 3° Celsius (1.8° - 5.4° Fahrenheit) warmer. At night the difference can be as large as 12° C (22° F). The heat island effect modifies the local climate, which could make a wider range of crops possible in cool climates and limit the usefulness of the USDA's regional climate map for any given growing site that is situated within a metropolitan area.

Besides the heat island effect, urban microclimates are created by the layout, configuration and size of buildings, which change sun exposure patterns and alter wind flow, the latter of which can trap air pollution in sheltered sites. The collective area of buildings and paved surfaces creates voids in the greenspace that do not provide the same environmental services that vegetated land provides, such as groundwater recharge and cooling from vegetation cover during hot weather, which can positively influence growing conditions.

The Social Environment

Although some urban food growing operations exist to produce a profit for the owners, many others are concerned with issues that lie well beyond economics. In urban areas, the close proximity of living and food growing spaces causes the two to become intertwined. In this context, some food growing operations take on cultural meanings and are used to address social issues. These meanings and issues include a desire to grow food using unconventional methods, the challenge of weaving food production into human living space, using urban agriculture as a part of sustainability initiatives, reducing dependency on external food sources and addressing community food insecurity concerns.

Whereas rural, large-scale commercial agriculture relies on established methods of planting, pest control, fertilizing and harvest, urban agriculture has a unique set of challenges that often cannot be addressed by industrialized or standardized agriculture methodologies. These challenges can become barriers to growers. Some barriers identified by home and community food gardeners from around the world included insecure access to appropriate land, inadequate gardening skills and a lack of social or cultural encouragement (Shutkin 2000, Subair & Siyana 2003, Kortwright 2007, Zahina-Ramos 2013). Access to land is essential for urban food growing but access alone does not guarantee suitability; land for food production must have unpolluted soils, appropriate light regime and irrigation water available. The acquisition of these resources in a heavily developed area often requires coordination with governing entities or neighbors. The need for local gardening skills and growing information, including encouragement from peers, is necessary to successfully cultivate crop plants and protect them from pests and diseases. These may be available in the community, but if disorganized, considerable effort would be needed to locate and access.

Just as food-growing conditions can vary widely across a metropolitan area, so can receptivity to the practice of growing food in cities. Some communities have embraced urban agriculture and have created land use and zoning laws to support the practice. In other communities, urban agriculture is considered to be a "Not In My Back Yard" (NIMBY) issue that is banned by local bylaws alongside harmful activities such as the dumping of toxic wastes. During the 20th century, urban development has eliminated agriculture (Szlanfucht 1999) in two ways: first, by paving over local agricultural land as development expands and second, by excluding food growing from within the built environment. In fact, development and farming have been perceived to be conflicting land use types (Smit & Nasr 1992, Martin & Marsden 1999). Agriculture is generally considered to be a rural activity and is physically/socially separated from living space in the United States, but this is not so in other parts of the world. Outside the United States, urban and home food production have a long tradition and residents embrace the integration of agriculture into city life. For example, in Kenya and Tanzania, two out of three urban families farm (Lee-Smith 2010). In Taiwan, half of the urban families belong to farming associations and there are more urban than rural farmers in Japan, the Netherlands and Chile (Smit & Nasr 1992).

In some cases, the paucity of urban food growers within an area can be attributed to psychological barriers, such as a lack of food growing knowledge or residents dislike for the idea of growing food (Zahina-Ramos 2013). In other cases, the barriers include limited access to growing space or legal issues. Nuisance laws are sometimes used to prohibit livestock and fowl production in developed areas and food gardens are sometimes banned because they are perceived to be associated with these other activities. Many deed-restricted and upper income communities prohibit food gardens through landscaping guidelines and policies. Some tracts of land are unavailable because of restricted legal access or fears of legal liability in the event of an accident or injury. In some cases these concerns have been justifiable as community gardens were left unkept or unmanaged.

Receptivity to urban food growing varies not just between communities, but also between demographic groups and these differences can lead to unequal levels of receptivity from neighborhood to neighborhood. One survey of urban residents' attitudes and perspectives about urban food growing found a statistically significant ($P<0.01$) positive relationship between age and the likelihood that someone was a food gardener (Zahina-Ramos 2013). The same study found a correlation between an individual's relationship status (single, dating, short-duration committed relationship and long-duration committed relationship) and the likelihood that a person practiced food gardening. Demographic groups that had the lowest percent of food gardeners were urban residents that had less than 4 years of college education, were under the age of 30 and who were in lower income brackets (< \$30k/year, 2013 USD) (Zahina-Ramos 2013). These results indicate that neighborhoods with dissimilar demographic characteristics will also have differences in the receptivity to and the practice of food growing.

Sustainability

In 1987, the United Nations' World Commission on Environment and Development defined sustainable development and placed environmental issues into the political arena (Brundtland & Khalid 1987). It also stressed the international nature of issues such as environmental limits to economic development, loss of natural habitat, poverty and food security. The 2005 World Summit on Social Development further advanced sustainable development goals and the concept that sustainability lies at the intersection of environmental, economic and social sectors (Kates, Parris & Leiserowitz 2005). The concept that sustainability was not solely an environmental or economic issue brought to light its interdisciplinary nature and the broad reach of stakeholders who are affected by it. Figure 1 represents the conceptual relationships between economic, social and environmental sectors of sustainability as three sides of an equilateral triangle.

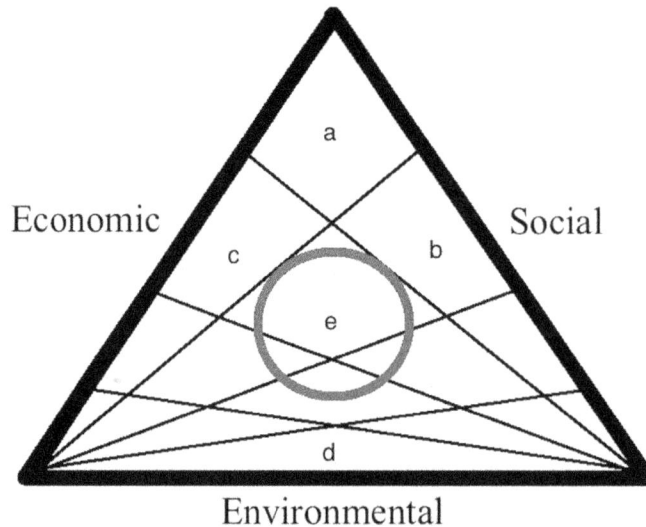

Figure 1. Conceptualized relationships between environmental, economic and social sectors of sustainability. The center circle represents sustainability interests that are balanced between the three sectors.

The representation of sustainability depicted in Figure 1 has two parts: the sides of the triangle and the interior space. The base of the triangle represents the environmental sector of sustainability, which is the foundation upon which human life depends. A healthy and sustainable environment provides ecological goods and services that are essential for human health and welfare, and are important economic and social drivers. For this reason, it has been placed at the base of the triangle. Economic and social sectors (both part of the social sciences) form the remaining two sides of the triangle, which converge at the triangle's apex- the point that is most distant from the environmental side of sustainability. The interior space of Figure 1 is broken by a series of lines that originate from the lower vertices and terminate at opposing sides. Any specific project, activity or issue that has a sustainability component (as a class, these will be referred to as a *sustainability interest*) may be thought of as lying somewhere along one of these interior lines. The location of a sustainability interest within the triangle visualizes the relative importance of environmental, social and economic sectors to that sustainability interest. A certain sustainability interest may be equally and closely related to social and economic sectors, but has little connection to environmental concerns; this sustainability interest would conceptually lie near apex of the triangle (near position "a" in the figure). Sustainability interests that are more heavily focused on social *or* economic sectors would fall near positions "b" or "c" in the figure, respectively. Some sustainability interests are almost entirely focused on the environmental sector and have only a minimal relationship with social and economic sectors; these sustainability interests would conceptually lie near position "d" in the figure. Sustainability interests that are more or less equally related to the three sectors would be found near position "e" in the figure.

One aspect of sustainability that is not represented in Figure 1 is resiliency. Systems that are sustainable have an ability to resist change or recover when a perturbation occurs. For example, intact natural ecosystems are able to recover following regularly occurring disturbances. Financially healthy businesses are able to endure normal economic variations and fluctuations. Socially sustainable communities can adjust and adapt to changes in governmental policy and politics. Resiliency is a necessary characteristic in order for a system to function over very long periods of time. It then follows that a sustainable food system is one that is able provide for those who depend upon it through the normal variations in environmental, economic and social conditions that can occur.

Sustainable Agriculture

Over the past several decades, concerns about urban sustainability have led to the development of new fields of inquiry that investigate and quantify the needs and resource consumption of population centers. Without exception, cities import many more resources than they produce within themselves. This has led to a situation where the lives of urban residents are dependent on the resources, materials and products that have been imported into the city and supplied to them, often through a commercial market system. The creation of a more sustainable city requires conditions in which it imports fewer commodities and exports less waste (Smit and Nasr 1992). Besides building materials, potable water and other resources that are needed to support life within a metropolitan area, food is one of the largest imports into cities (Smit and Nasr 1992) because it must be consumed in certain quantities and varieties to maintain human health. Since food is such a large part of the bulk that is transported into cities, urban food production can play a key role in sustainability and resource conservation.

The term "sustainable agriculture" can mean different things. To the economist, it means that the operation is generating a sufficient profit to ensure business longevity. Economic sustainability seeks to meet the following criteria: 1) the farming operation can be sustained through time without degrading the land, 2) it replenishes the crops or livestock, 3) it enables the operator to continue farming and 4) the operation is profitable (Macher 1999). To the social scientist, sustainable agriculture addresses the issues of human health and food insecurity. It may also include preservation of a family's or farmer's heritage. To the environmentalist, sustainable agriculture means that the operation is not negatively impacting natural resources such as soil and water. More recently, additional modifiers and clarifications have been added to what defines a sustainable food production operation. To be sustainable, the agricultural operation must not:

- Pollute the environment, including downstream areas
- Cause ecological impacts (change) either at the food production site or elsewhere
- Cause secondary impacts, such as greenhouse gas emissions, which are often unseen

This more robust definition of sustainable agriculture is more difficult to measure, but increasingly important as the world's population continues to rise at unprecedented rates and the need to increase food production is a critical concern. Below, three primary objectives of sustainable agriculture are described. These objectives form the foundation upon which agricultural sustainabilty will be viewed throughout the remainder of this book.

Objective 1: Meet Present and Future Needs

Rees (1992), Rees & Wackernagel (1996), and Costanza *et al.* (1997) put forth important definitions of what it means to be sustainable. These authors described sustainability as the ability for a community to meet its present needs without impacting the ability of future generations to meet their needs. This criteria for sustainability requires that resources be used in such a way that they are not depleted or degraded through time. It also assumes that there is some degree of resiliency to allow the community's needs to be met during periods of normal disturbances.

As one grows food, resources are consumed and supplies are used (inputs), plants uptake nutrients and water (internal processes), and products are harvested and materials leave the site (outputs). Some of the materials that can leave a food-growing site are immediately identifiable (and quantifiable) while others are not visible. For example, when fertilizers are applied to a growing site, plants take some up some of the applied nutrients, some may be lost through surface runoff during rainstorms and some may leach into the groundwater. If a growing operation uses materials or fertilizers that are available from limited or non-renewable sources, then that agricultural practice is endangering the ability of future generations to use that same resource. This would be regarded as an unsustainable practice. In contrast, an agricultural operation that does not exhaust the resources it uses ensures the long-term viability of the food system. This also allows the farmer to produce over a long span of time and the operation can pass through successive generations. This is an extremely desirable situation from economic, environmental and social perspectives.

Any food producing operation that is able to achieve a sustainable condition is extremely valuable because of its vitality and viability in the long term.

Although this book is preoccupied with measuring the benefits of urban agriculture, there is less emphasis placed on the social aspect of sustainability. This does not mean it is less important. Social sustainability is often poorly defined, complex and difficult to quantitatively measure. Where possible, the social aspect of sustainability will be addressed. But, much of the emphasis of this book will be on economic and environmental sustainability since these are more suited to quantitative methods.

With respect to the social sector of sustainability on a larger scale, the collective output from agricultural operations must provide for an adequate human diet. If it does not, then those who rely on these operations will suffer from health problems. This issue is not necessarily applicable to output from a single farm, but to the collective products that are grown to supply a particular population. As the world's commercial markets become increasing integrated, the regional or local crops produced in an area become less relevant since any deficit in variety can be made up through imports. However, long-distance transport of food creates problems that are incompatible with sustainability principles. Hence, local and regional production of an appropriate suite of crops that support a healthy human diet is within the domain of social sustainability. More on this topic will be presented in later chapters.

Lastly, with respect to the relationship between social sustainability and sustainable agricultural systems, the agriculture must continue to provide food during times of fluctuating conditions and perturbations- in other words it must be resilient in the face of change or disturbance. Examples of change or disturbance include economic hardship or recessions, climate change, rare climatic events and changing social conditions (e.g., trends). Urban agriculture, by nature, is much more highly adaptable and resilient than commercial rural agriculture since it is often detached from the commercial market system, reflects local conditions and is readily adaptable to change.

Objective 2: Resource Stewardship

The renewable environmental resources that agriculture relies upon include water, nutrients, soil and local ecosystems, which contain beneficial organisms such as pollinators. There are other non-environmental renewable environmental resources that are important to agricultural production; these include sunlight, labor, working capital and a demand for agricultural products. Orr (2002) and Edwards (2010) noted out that many long-lived societies demonstrate an understanding of resource consumption limits and resource stewardship. The recognition of resource limitations are respected at the cultural level and passed through generations, which allow for the resource to be protected rather than exhausted. The fact that some essential resources are limited implies that there is also a limit to the amount of agriculture a plot of land can support before its resources become irreparably harmed. This concept is referred to as the land's carrying capacity, which will be discussed in more detail later.

To illustrate the importance of using and protecting a renewable resource, consider the example of soil. On traditional farms, soil organic matter and nutrients from natural sources are essential resources to grow healthy crop plants. In order for cropland to remain productive and to be sustained in the long-term, the soil's organic matter and nutrients must be continually replenished (Peoples, Herridge and Ladha 1995). In fact, agricultural practices that do not maintain soil organic matter and fertility will eventually leave the operation with shrinking productivity and value. One way that renewable resources can be protected is by recycling and reusing organic materials. Plant-based waste and compostable materials can be turned into valuable soil amendments that recycle nutrients and organic matter that would be lost if disposed in a landfill.

Objective 3: Environmental Protection

Natural ecosystems provide a large suite of ecological goods and services that humans depend upon. In some cases, these goods and services provide direct economic benefit and in other cases these benefits are manifested as essential services that support human life in secondary ways. Humanity and its agricultural systems are dependent on natural systems and the goods and services they provide, such as clean air, water and soil. Sustainable agricultural systems must protect these ecological goods and services by preserving environmental quality, ecosystems and biodiversity.

Orr (2002) and Edwards (2010) described the sustainability of agricultural systems in terms of the interconnectedness between human and natural systems. Their discussion of sustainability referenced successful ancient agricultural systems as models, noting that those systems thrived as part of a healthy, unpolluted and intact environment in which the community lived. However, those models are based on human population levels that are within the area's carrying capacity. What about places where the land's carrying capacity has been exceeded and the population is only sustained by products and resources that have been imported from somewhere else? That is the condition in most cities today. Because urban resident are supported by imported resources (especially, food), the ecological impacts resulting from extracting and supplying those resources are deferred to locations outside the city. There are concerns about the degree that today's society can rely on large-scale, long-distance transportation of food, especially with respect to fossil fuel consumption and the resulting greenhouse gas emissions (Peters *et al.* 2009a).

Ghosh (2010) examined suburban domestic gardens in Australia and found that suburban backyard vegetable gardens can provide multiple sustainability benefits including local food production, recycling of organic wastes through composting, lower urban carbon footprint, water and energy conservation, improved public health and better social connections. Community gardens and residential food gardens can also contribute to storage of atmospheric carbon as biomass, urban biodiversity and other important ecological functions (Okvat & Zautra 2011).

Some food growing methods may appear to be "sustainable" on the surface or from a certain perspective, but may not be when the full range of costs are examined. This is illustrated in the following example: a community garden is being established on a vacant urban lot and the garden's leadership decide to invest in raised beds, which will be constructed from a rot-resistant wood for durability. Rather than using the local soil, which has a high percent of sand, they decide to purchase organic garden soil so that they will have better moisture retention. They install an irrigation system out of PVC, which is connected to a groundwater well and electric pump. The garden only uses organic methods and will be planting a number of fast growing and fast reproducing non-native species that produce products that are known to repel pests.

In this example, the community garden's leaders consider their operation to be "sustainable" because they are producing local food using organic methods. But, a closer look will show otherwise. First, the wood used to create the raised beds was not produced from a sustainably managed forest. Because of this, the community garden's choice of building materials may be contributing to habitat destruction. Additionally, the wood has been shipped from a forest that was thousands of kilometers away, so it carries a non-renewable energy and CO_2 cost not accounted for by the community garden leaders. With respect to the irrigation system, it was made from manufacturing processes that create pollution and consume non-renewable resources; the materials were then shipped thousands of kilometers to the consumer. The organic soil used by the garden was composed of peat, which was mined from a drained wetland in another region. This organic soil was obtained by a process that destroyed natural habitat and was shipped over a thousand kilometers. Lastly, the non-native species that were planted for pest control have been found to be highly invasive when cultivated outside of their native range. If these species escape and invade local natural habitats, they may eventually cost millions of dollars to control. There are other practices that are not sustainable too. Over-fertilization can cause polluted runoff to enter nearby streams or groundwater after a rainstorm. Pesticides from natural sources that are accepted for use in organic agriculture may be used in ways that harm the environment. So, when summarizing the broader range of environmental costs associated with this community garden, one has to conclude that it is probably not a sustainable operation. The negative environmental impacts of this one garden may not be significant, but the cumulative impacts of hundreds of community gardens like this one do become significant.

The concept of ecological farming has been developing over the past decades as a response to the concerns of sustainability, preserving ecological functions and environmentally friendly management of food growing spaces. Ecological farming seeks to preserve and restore many of the ecological goods, services and functions that are found in natural systems while producing an adequate and sustainably grown suite of crops. Some of the ecological functions include groundwater recharge, prevention of soil erosion, building of soil organic matter content, the capture and cycling of plant nutrients, assistance in the control of damaging agricultural pests by supporting natural predator/prey interactions and increasing (native species) biodiversity. Ecological farming seeks to more closely approximate the function and structure of a natural ecosystem than other agricultural methods, such as organic farming. To date, ecological farming is more concept than established practice. Much research work lies ahead to document growing methodologies and to examine productivity, ecological and economic data to determine how food-growing practices can be supportive of a healthy environment, rather than a liability.

Based on the above discussion, the basic principles of sustainable agricultural systems include:

- Use of resources without depleting them
- Resource consumption limits are recognized and respected
- The environment is protected (remains healthy, unpolluted and intact) while the agricultural activity is being conducted
- The products from sustainable agriculture must meet the dietary needs of the population it supports; this assumes that what is produced supports a healthy human diet with the appropriate variety of food types
- The sustainable food growing operations are economically viable

Chapter 5. Environmental Conditions and Agriculture

Besides the actual sowing of plants, there is perhaps no factor that influences agricultural production more than the environmental conditions under which the plants grow. For this reason, a great deal of agricultural and horticultural research has been focused on:

- Understanding the optimal conditions required for crop plant growth and production
- Creating crop planting schedules for specific climates
- Breeding plants to tolerate sub-optimal environmental conditions
- Designing climate-controlled agricultural production systems (e.g., greenhouses, hydroponics, etc.)

This chapter reviews the ways that environmental conditions can influence crop production, describes some aspects that are important to consider and provides examples of how environmental conditions can be quantitatively characterized.

Plant Productivity

The practice of growing food is subject to certain risks and uncertainties arising from environmental conditions. Nature has a significant influence on farming (when conducted outdoors) and forces beyond the control of the individual farmer largely determine the quantity and value of agricultural products (Debertin 2012). Agricultural products are measured and valued in different and often conflicting ways. How any plant product is valued is a function of market and social factors. Agricultural crop value is often expressed as a measure of productivity and this is normally focused on a specific plant product. In ecological research, plant productivity measures are used to understand how much of something has been produced over a period of time and under a set of known environmental conditions. Without environmental data, the context of the production numbers is lost and the results are of limited use. This is because many different factors can regulate or influence plant growth and reproduction cycles.

Farmers know that they can get the soil amendments and irrigation right, but weather is the master control that determines if there will be a productive year or not. The greatest agricultural failures occur when there is discord between the crop plant growth preferences and the ambient environmental conditions. Conversely, the best harvest arises from a growing season that has weather conditions that are most compatible with crop plant preferences.

Measuring and characterizing environmental conditions are not the same thing. For example, one *measures* an environmental parameter when taking air temperature readings. One *characterizes* environmental conditions when those measurements are analyzed using some statistical method, such as calculating the average daily maximum temperature over a certain period of time (e.g., the month of July). If an investigator lives in an area where there is a nearby environmental monitoring station that measures a robust set of parameters (such as temperature, rainfall, relative humidity, wind speed, pan evaporation, solar radiation, etc.), then the local environmental conditions can be well documented and characterized. Often, these data can be acquired for little or no cost from government agencies.

When an investigator does not live near an environmental monitoring station, then it may be necessary to take measurements. This can be time-consuming and it can be expensive to collect data according to strict measurement standards that guarantee accuracy. If one must collect their own data, then it is necessary to prioritize the importance of different environmental parameters and estimate the cost for acquiring the data when deciding what to measure. In some cases, there may be parameters that are critically important, but these lie beyond the affordability threshold of the investigator.

Because weather conditions can vary between seasons and from year-to-year, long-term comparisons between crop production and growing season environmental conditions are enormously beneficial. Seed catalog and crop growing recommendations published by agricultural universities are based on a generalized and simplified set of guidelines that one should use under average regional conditions. However in urban areas, each growing site is unique and those guidelines can accumulate a long list of caveats and footnotes once the grower begins to gain experience. Many of these caveats are the result of local landscape variability that can influence local climatic conditions.

The characterization of environmental conditions can be tricky. For example, researchers report average temperature or rainfall over a season, but these statistical metrics have only limited relevance to plant productivity studies. The reason for this is complex and some the issues are:

- Generally, plant productivity is optimal when environmental conditions are within a certain range for a certain length of time. When conditions fall outside of this range, productivity can decline. The longer conditions stay outside of the optimal range for plant growth, the greater will be the loss of productivity.

- Crop plants originate from a wide range of environments around Earth. Because of this, different crop plants may have different (but overlapping) optimal growth ranges. When growing a variety of crops together, each species may each respond differently to a particular set of environmental conditions.

- Excursions outside of the optimal range of growing conditions affect plant growth differently. For example, extremely high temperatures affect plant respiration, can cause wilting and damage leaves but may not be deadly. Some of these impacts can be mitigated with adequate soil moisture and, once temperatures return to a more normal range, the plant can recover. When extremely cold temperatures occur (e.g., freezes), exposed parts of the plant may be permanently damaged and in some cases, the plant will die.

- Many crop plants do not grow best when environmental conditions are static. For example, some long-season crop plants from temperate climates grow best when temperatures are cool during the seedling state, temperatures are warm during the main growth period and cool temperatures return as the plant nears the end of its lifespan.

Given the complex nature of plant productivity as described above, a more comprehensive and refined examination of environmental conditions is warranted. More of this will be described below. Some of the main environmental and climatic factors that influence the productivity of urban agriculture are:

- Climate, both local and regional
- Growing Season Temperatures
- Growing Season Rainfall
- Extreme climatic events
- Soil
- Exposure
- Pests

Below are descriptions of these factors and how they are related to plant growth and productivity.

Climate

Climate is defined as the long-term weather conditions of an area. It is usually expressed as a set of statistics (mean, maximum, minimum, etc.) that are calculated from measured parameters over a 30-year (or more) period. These parameters include temperature, precipitation, humidity, wind speed and atmospheric pressure. Weather is defined as the short-term conditions (e.g., daily or weekly), which can also impact food-growing efforts. Local climate and weather can have a profound effect on agricultural productivity. For this reason, it is essential to understand where the urban food production is or will take place. Several aspects of climate that are important to consider include the growing season duration, temperature regimes, moisture availability and other limiting factors.

On a global scale, climate is predominantly governed by latitude and elevation (Figure 2). Local physical geographical features can modify the climatic conditions at a specific area, creating a complex landscape that can reduce or magnify the effects of exposure, evaporation, temperature and other environmental factors. A more detailed discussion on the subject of complex landscapes is included in a later section.

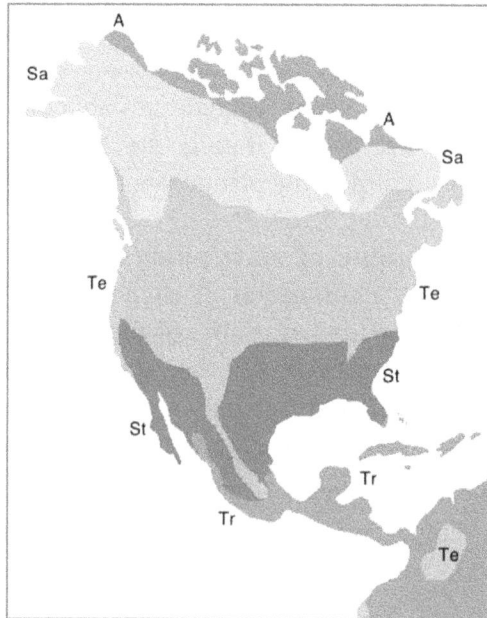

Figure 2. Generalized climatic zones associated with latitudinal changes in the Americas: A= arctic tundra, Sa= subarctic, Te = temperate, St = subtropical, Tr = tropical. The southward trending temperate zone lobe into Central America and the temperate climate inclusion in northwestern South America are associated with high elevations.

Growing Season Duration

When studying agricultural productivity, the period of time during the year when conditions allow for the (usually optimal) growth of crops is referred to as the growing season. How the growing season is defined varies from culture to culture. In

North America, the growing season is defined by frost events, which damage or kill food plants. The growing season is regarded as the time between the average dates of the last spring frost and the first autumn frost (in areas where frost occurs every year). Almanacs published in the United States and Canada typically report these average dates. Because the Earth's climate has been gradually warming over the 20th century and climate is typically based on 30-years of weather statistics, the length of the growing season, according to North American definitions, has gradually increased.

The continental European definition of the growing season is based on the annual average number of days where the mean 24-hour temperature does not fall below 5°C (6°C in some areas). Unlike the North American definition, this growing season is based on growing conditions rather than risk of plant damage. England uses a variation of the 5°C criterion used in Europe, requiring five consecutive days at or above 5°C to mark the beginning of the growing season and five consecutive days below to mark the end of the growing season.

In other parts of the world, where cold temperatures may not occur regularly, the growing season may be defined (either beginning or ending) by the onset of the rainy or dry season. This occurs in some frost-free parts of North America and in the Caribbean Basin. Where climatic conditions are relatively stable throughout the year, the growing season is all year.

Climate and Latitude

Temperature, day length and sun intensity vary along the Earth's latitudinal gradient. The higher latitudes typically have a greater range of seasonal temperatures, day lengths and sun intensity than is experienced in equatorial regions (Table 4). Although temperature is a well-known controlling factor in agricultural production, two lesser-appreciated differences between higher and lower latitudes are light quality (intensity) and the length of day/night (photoperiodism). With respect to the latter, some important agricultural crops are obligate or facultative with respect to short-days (long-nights); these include cotton and rice. Obligate or facultative long-day plants include oat, pea, barley, wheat and lettuce (Hamner & Bonner 1938). Some crops, such as cucumbers and tomatoes, are neutral with respect to day/night length.

Climate and Elevation

Just as latitude can alter growing conditions, so can changes in elevation (Table 4). As one goes higher in elevation, such as up a mountain slope, environmental conditions change. Climatic variables along elevation gradients influence plant growth, the length of the growing season and exposure to adverse weather. One noticeable change is air temperature, which decreases at a constant rate as one goes higher in elevation- a phenomenon known as the lapse rate. This phenomenon causes higher-elevation regions to have cooler climates than lower elevation regions at the same latitude, provided other influencing factors are comparable.

Table 4. Average growing season durations for selected cities.

City, State	Mean Growing Season (days)[1]	Latitude[2]	Mean Elevation[2] (m)	Comments
Miami, Florida	365	25° 46'	2 (6 ft.)	On Atlantic Ocean
Savannah, Georgia	268	32° 1'	15 (49 ft.)	On Atlantic Ocean
Chicago, Illinois	186	41° 50'	181 (594 ft.)	On Lake Michigan
Green Bay, Wisconsin	150	44° 30'	177 (581 ft.)	On Lake Michigan
Juneau, Alaska	148	58° 2'	17 (56 ft.)	On Gastineau Channel
Denver, Colorado	156	39° 45'	1,648 (5,410 ft.)	Mountainous
Provo, Utah	167	40° 14'	1,387 (4,551)	Mountainous
Bismarck, North Dakota	129	46° 48'	514 (1,686 ft.)	-
Rapid City, South Dakota	140	44° 0'	976 (3,202 ft.)	-
Lansing, Michigan	145	42° 44'	262 (860 ft.)	-
Cincinnati, Ohio	192	36° 6'	147 (482 ft.)	-
Nashville, Tennessee	204	36° 1'	182 (597 ft.)	-
Athens, Georgia	227	33° 57'	194 (636 ft.)	-
Tallahassee, Florida	239	30° 27'	62 (203 ft.)	-
San Antonia, Texas	269	29° 25'	198 (650 ft.)	-

[1]Source: Yankee Publishing 2016
[2]Source: Wikipedia

Other environmental conditions that may change with elevation are light intensity and wind exposure. These differences are most pronounced when comparing summit conditions to those in sheltered valleys below. Some north-facing slopes of large hills or mountains in the northern hemisphere, as well as valleys in high-relief areas, may be entirely unsuitable for crop production because of insufficient sunlight exposure.

Temperatures

Local temperature regimes are often described as a suite of statistics that have been calculated for a given span of time. The most commonly reported weather statistics are for a 30-year average or longer (climate statistic), a year (e.g., annual average) or for a season (e.g., average daily high temperature). When inquiring about the temperature statistics of an area, care must be taken to select those parameters that will best address the issue of concern. Some temperature statistics that are of interest to agriculture operations are described below Table 5 and are described in the following sections.

Table 5. Temperature statistics that are important to food growing operations.

Temperature Statistic	Relevance
Average	Mean temperature value over a period of record
Range	Can be used to determine if the normal range of temperatures is compatible with optimal growing conditions for selected crop plants
Minimum	Can be used to determine if minimum temperature exposure may cause crop damage
Maximum	Can be used to determine if maximum temperature exposure may cause crop damage
Extremes	Frequency of extreme events is inversely related to crop productivity

Average Temperature

Some meteorologists often refer to "normal temperature" and "average temperature", but these are meant to infer the same thing- a calculation of the mean temperature that has been measured from a specific weather station or site. The average temperature can be based on any specified time period (day, week, month, year or decade) or for extremes (e.g., the daily high or daily low). As discussed earlier, average temperatures are only a partial characterization of the local conditions that are important for those who grow food crops, as the statistic lacks information about the occurrence and frequency of damaging extreme conditions that may be short-lived.

Temperature range

The temperature range is a statistic that describes the minimum and maximum temperature that occurred over a specified period of time. For example, on a day when the maximum recorded temperature was $26°C$ and the minimum temperature was $12°C$, the temperature range was $14°C$. The range statistic can be calculated for any specified time period (day, week, month, year or decade).

The temperature range can be useful to those who are growing crop plants that prefer to have a fairly constant temperature regime or crops that prefer fluctuating temperatures. For example, some crops originate from tropical lowlands where temperatures are relatively stable throughout the day and throughout the year. These crop plants grow best when temperatures are warm and relatively stable. Other crop plants, such as those that originated from higher latitudes or higher elevations, prefer warm days and cool nights, and some are even tolerant of light frost. These crops prefer a wider temperature range than those from the lowland tropical region.

Temperature range measurements provide growers with valuable information about the suitability of local growing conditions for a specific crop plant of interest.

Temperature Extremes

Excessively hot or cold temperatures stress crop plants and can cause injury or death. Extreme temperature events can be short-lived (e.g., an overnight frost) or long in duration (e.g., heat wave). Extreme cold temperatures affect plants differently than extreme hot temperatures do. Freezing or near-freezing temperatures can cause leaf or plant death, the result of damage to cells and vascular tissues during the freezing-thawing process. Extreme heat tends to cause wilting and desiccation of plant tissues, which are especially damaging if insufficient moisture (soil or air) is present. As with temperature range, crop plant tolerance to cold and heat is related to the environment it originated from. Exception can be found in hybridized varieties that have been selected for tolerance to certain adverse conditions.

The quality of some crop varieties is improved when temperatures exceed the optimal range. For example, the flavor of parsnips, leeks and some cole crops (e.g., cabbage, kale, Brussels sprouts) are improved after they have experienced freezing temperatures. Even though productivity may be affected, the better tasting crop may demand more in the marketplace.

Wind

Wind plays an important role in maintaining air quality within urban food growing sites. Many crop plants grow best in an area that has at least some degree of airflow, which ensures an adequate CO_2 supply for photosynthesis and prevents a buildup of air pollution. Wind also helps to regulate and homogenize temperature conditions at a site.

Excessively strong winds or persistent wind stress can stunt plant growth and damage plants. Severe wind gusts can topple plants and tear leaves, flowers and fruit from plants. Strong winds combined with cold or hot temperatures can cause serious harm to crop plants. In areas that are prone to excessive winds or persistent wind stress, windbreaks are usually planted or installed to reduce the potential for wind-related damage.

Moisture Availability

All agricultural plants need an adequate supply of water to grow, reach maturity and produce a harvestable crop. Moisture availability is defined as the amount of water that is available for crop plants to grow. For rooted plants in gardens and fields, soil moisture is the primary water source since most crop plants take in soil moisture directly through roots. However, humidity levels also influence the amount of water that can be lost through leaf pores.

Factors that can influence moisture availability include:

- Rainfall; both magnitude and temporal aspects affect soil moisture and humidity
- Evaporation potential, which is the ability of water to evaporate from soil, leaf surfaces and to be lost through leaf pores
- Soil type; some soil types have a high water-holding capacity while others (e.g., sand) are poor at holding soil moisture
- Topography; low-lying landscapes, such as valley floors or floodplains, often have higher soil moisture contents because runoff following a rainstorm tends to accumulate in these areas
- Geology; underlying geology may influence soil moisture- subsurface confining soil or rock layers can create a perched water table and saturated soil conditions

Insufficient moisture availability can have catastrophic effects on some crop plants, particularly those with higher growth or respiration rates. When there is inadequate moisture to maintain plant cell turgidity, the plant wilts because leaf and stem cells are unable to maintain water content. This halts plant growth and can damage fast-growing tissues (such as buds and immature leaves). If drought conditions persist or become worse, the plant will reach the permanent wilting point where cells can no longer recover turgidity. Once this happens, these tissues become damaged and necrotic. If the permanent wilting point is maintained for an extended period of time, the crop plant may never be able to recover and will die.

With respect to agriculture, moisture availability is usually compared with moisture demand. Moisture demand is the amount of water that is needed to successfully grow a crop to harvest, which varies according to the crop type. When moisture availability is less than the moisture demand, crop production will decline unless irrigation is supplied. Moisture demand is also influenced by the local climatic conditions. For example, the amount of water needed to grow corn in Colorado is 26% more than the amount needed in the eastern United States (Macher 1999). This discrepancy can be attributed to the lower humidity and rainfall conditions found in Colorado.

A drought is defined as a period of below average precipitation, which causes depletion of soil moisture and groundwater supplies. Droughts may be seasonal in duration or can last for many years. Some regions of the world experience a dry season each year, which can exasperate the effects of a drought.

The discrepancy between moisture availability and crop moisture demand can bring about a phenomenon called a "hidden drought". A hidden drought occurs when there appears to be sufficient rainfall but agricultural crops are not receiving enough moisture for proper growth. This can occur in arid or semi-arid regions that have a high evaporation potential (evapotranspiration rate, which is the combined effect of evaporation and transpiration of water by plants), which can quickly deplete soil moisture and cause plant stress. In more humid regions, a hidden drought can occur when brief rainfall events do not produce enough precipitation to increase soil moisture. In this case, most rainfall is left on plant and soil surfaces, evaporating after the shower has passed. Soil moisture is not recharged by these brief periods of rainfall and is gradually depleted as plant roots continue to extract water for photosynthesis.

Excessive soil moisture can be detrimental to food crop production and in some cases may be more serious than drought. Prolonged, excessive rainfall patterns can cause crops to become stressed and susceptible to disease. Low-lying areas may be flooded during extreme rainfall events and the crop plants may be killed in a matter of days. Careful site planning and investigation of potential areas of flooding or ponding of runoff are necessary to avoid some of the problems associated with excessive soil moisture.

Soil

For in-ground agricultural operations, soil condition and composition are important factors in productivity. Because of this, much research, governmental policy and public programs have been focused on soil and its conservation. Although these efforts have mostly been concerned with commercial rural agricultural operations, soil is no less important to the urban agriculturalist. Agricultural soils are a valuable resource that must be monitored, managed and protected.

Soils are classified using a taxonomy system that is based on physical and chemical characteristics. Soil moisture regimes and temperature regimes are also used to classify soils. Although soil classification is an area of major interest to large-scale commercial operations, it becomes less important to small-scale urban food growers because of their ability to manage and manipulate the relatively small area of land they grow in. In many urban settings, agriculture production is often conducted on previously- disturbed land that has lost its original soil characteristics. Soils can be of natural origin (such as those defined in the soil taxonomy system), of artificial origin (manufactured soil and soil-like mediums), or a mixture of both. Manufactured soils, soil mediums and soil mixtures that have been created for agricultural uses have often been made from recycling waste products from another process and are sterile (free of pests and diseases).

Soil is composed of minerals (most are highly weathered), organic matter in various states of decay, liquids (mostly water), gasses and a rich community of organisms. In essence, soil is an ecosystem that has living and non-living components that interact and influence the soil environment. The soil's physical and chemical characteristics can be a limitation on agricultural productivity.

Many informational sources often refer to "soil condition," but one must use caution when determining if this term is appropriately defined for agricultural usage. A soil's condition is usually defined from a specific perspective and can be relative. For example, if erosion is carrying away topsoil from a farm field, that soil is considered to be in a degraded state because it is gradually losing its value for agricultural production. If native forestland is cleared for agricultural production and the soil is found to be pure sand, the soil is considered to be poor (from the agricultural productivity perspective) even though ecologists would consider the soil to be appropriate and natural for the native ecosystem it supported. Because these differences in perspectives often lead to conflicting soil appraisals, it is important to understand why they exist and where they come from.

Soil ecology can be complex. Innumerable microorganisms live in soil and perform important roles in nutrient cycling, decomposing organic matter and plant absorption of water and minerals. Soil macroorganisms (e.g., beetles, worms, ants) often bridge the interface between the subsurface and surface environments, increasing soil aeration, water infiltration and the vertical movement of dissolved materials. Soil organisms may be agriculturally beneficial, benign or pathogenic/pests and they are sensitive to changes in the soil environment. When certain practices are used that alter soil chemistry or hydrology, the soil ecology will also be changed.

Soils can be naturally fertile (i.e., contain an adequate supply of plant nutrients) or infertile (sometimes referred to as "sterile"). Soils that are rich in organic matter, have some clay and/or are derived from parent material that contains essential plant nutrients are naturally fertile. Soil fertility can be increased by the addition of organic matter, clay and organically- or inorganically-sourced nutrients. Collectively, these are referred to as soil amendments. Other soil amendments are used to change the soil's pH, increase water-holding capacity and to decrease soil density, which retards water movement through the soil and gas exchange with the atmosphere.

Besides fertility, another aspect of soils that is important to agricultural operations is the water holding capacity. Soils that are best able to hold and store moisture between rainfall (or irrigation) events without creating saturated conditions are classified as having a high water holding capacity. These include soils with a high amount of organic matter and/or fine-textured materials, such as clay or silt. Soils that are good at retaining moisture between rain or irrigation events are also better able to hold dissolved nutrients and minerals (rather than lose them to leaching) and supply these to plant roots.

Soil conservation efforts are focused on reducing or eliminating soil erosion and maintaining the soil's ecology, fertility and water holding capacity. Soil conservation is achieved through several methods, some of which include:

- Reducing runoff from the cultivated area
- Reducing soil disturbance (e.g., tilling, compaction, etc.)
- Replenishing soil organic matter and fertility by using cover crops or amendments
- Installing windbreaks to prevent soil wind erosion

Agricultural Pests

Every region has its own set of climatic conditions that support an array of agricultural pests, some of which are native and some have been introduced from other areas. Agricultural pests can be organized into specific groups: animals, plants and diseases. With respect to urban agriculture, the approach used to control these pests is often at odds with that used by large-scale rural agricultural operations. Large-scale operations often grow a few types of crops, which is a perfect environment for a pest to thrive. For them, large-scale pest eradication is the preferred method of control. Urban and near-urban food growing operations typically grow a variety of crop types, each of which may have different susceptibilities to pests. Eradication of pests may not be possible or practical in urban settings as these can reside on a neighboring parcel of land and can readily repopulate a site from which they have been extirpated. For urban and near-urban food growers, pest control and management is a better and more realistic option than eradication.

Complex Landscapes

Complex landscapes modify local climatic conditions on relatively small scales and create a mosaic of conditions that require site-by-site assessment to best determine what can be grown and how. Complex landscapes can arise from three variables that are found in the urban landscape: (1) the physical urban form, which arises from development patters; (2) ecological gradients and patterns of greenspace; and (3) environmental gradients and resources. These are further discussed below.

Because urban landscapes are complex in form, the urban food grower can be faced with the daunting reality that crop-growing advice available from the agricultural extension agent or studies conducted on rural farms are often inapplicable to the urban landscape. Indeed, growing advice from someone across town whose garden is in a sunny, open site may not be useful to the gardener in the city center who is surrounded by tall buildings.

Urban Development Patterns and Forms

At the local level, urban landscapes are highly variable and complex in relief, mostly due to building and landscaping forms that reflect different development styles and objectives. This complexity of form and relief creates a wide variety of growing site areas, shapes and climate modifiers that require urban food growers to assess and consider (Figure 3). Some of these modifiers are:

- Reduced wind and air flow across the landscape, as compared to non-forested rural areas (site A air flow is restricted by buildings all around, site B is open from north to south and air flow at site C is entirely blocked from the west)
- Reduced sun exposure and shade caused by buildings, opaque fences and trees (site C has east exposure, site D has west exposure and site E has little sun exposure)
- Modified soil moisture regimes resulting from rainfall capture, diversion and drainage features that transport runoff into sewers or swales, causing some areas to be drier and others to be wetter
- Temperature microclimates around buildings and certain types of infrastructures, such as heat-absorbing walls, ventilation system outflows and large conduit/pipe networks that may modify temperatures below or above ground
- The urban heat island effect modifies the diurnal temperature regime as compared to undeveloped rural areas

Ecological Gradients and Patterns in Cities

Cities usually contain a patchwork of different greenspace types that have been superimposed on ecological gradients, both of which can be important to urban food growing efforts (Figure 4). The most obvious ecological gradient runs from the city center, through the suburbs and to the countryside. Urban areas also contain a patchwork of greenspace, some of which is on undeveloped land and some in contained in local parks and preserves. Preserves and undeveloped land may include relic natural habitats whereas parks and developed land are often planted with vegetation that has aesthetic, rather than ecological value.

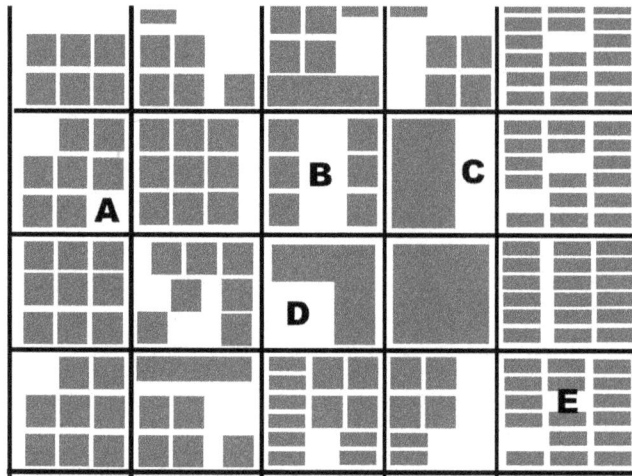

Figure 3. Representation of the complex urban environment and extreme heterogeneity of growing conditions; black lines indicate roads, gray blocks indicate buildings and white areas indicate greenspace, north is up.

Figure 4. Digital aerial photo of the Miami, Florida metropolitan area c. 1984 showing a gradient of built, planted and natural land cover from the urban core (right) to the urban fringe (left) gradient; image source: University of Florida Digital Aerial Photo Archive Collection.

These ecological gradients and patterns of urban greenspace can influence urban agriculture in several ways. These include:

- Differing physical conditions (air and soil temperatures, moisture regimes and wind and sun exposure) between habitat types
- Different organisms, some beneficial (e.g., pollinators) and some destructive (e.g., agricultural pests) occur in different concentrations along ecological gradients and in urban habitats
- Greater biodiversity is associated with more structurally complex ecosystems because there are more ecological niches; higher local biodiversity (in urban and near-urban areas) can provide more ecological goods and services that urban agriculture can benefit from

Environmental Gradients and Resources

Environmental gradients exist in cities that significantly affect the suitability of a site for growing food. These gradients include:

- Climatic gradients that occur from coastlines to interior areas; large water bodies have a moderating effect on temperatures and, sometimes, moisture levels
- Soil nutrient and organic matter gradients that occur in urbanized areas that are situated on drained wetland sites or in historic floodplains
- Pollution gradients associated with point-source emissions or outfalls, or non-point source emissions that are associated with urban density
- Moisture gradients associated with elevation changes or geological features

Chapter 6. The Economics of Urban Food Production

In the most basic sense, the field of economics is preoccupied with the interdependent linkages between production, distribution and consumption of products and services. At the granular scale, how individual elements interact and produce outcomes is referred to a microeconomics. These elements include individuals (e.g., a person, household or business) and their participation in markets. At the broader perspective, analysis of an entire economy is referred to as macroeconomics. Macroeconomics examines influencing policies, resources (e.g., capital, labor force), inflation and economic growth/decline. Because food is such a basic human necessity, the production, distribution and consumption of food products represent significant and important economic forces. The economic characteristics of the different types of agriculture can be remarkably different, making comparisons between them tenuous at best. In this chapter, the microeconomic characteristics of urban agricultural operations are explored and contrasted. By looking at the fundamental differences between the economic workings of different agricultural systems, one will gain a greater understanding of their individual roles in society.

Recall that microeconomics is concerned with *individuals* and their participation in *markets*. For the purposes of this discussion (and throughout this book), the individual will be defined as the food-growing operation and the market will be defined as any of the ways in which the crops are exchanged between other parties. The urban food growing operation (individual) may indeed be one person or it may be an entity, such as an urban farm that employs many people. Of particular interest to the individual is the identification and quantification of input (which come at a cost) and output (which can generate income through barter or sale) values so that profitability (output value exceed input value) can be determined. Inputs can be monetary expenditures to obtain goods or services, but they can also be unpaid labor and intangible things like emotional commitment to success (which are often difficult to put monetary values on). Similarly, outputs can be financial (dollars), non-financial (trade) or intangible (enjoyment of the food). Although inputs and outputs may be shared by different food growing operations, they will usually have different levels of importance and meanings.

Food-growing activities that are conducted within urban and near-urban settings can be divided into non-commercial (not for profit) and commercial (for-profit) operations. Non-commercial operations include family-owned subsistence farms, non-profit farms, community gardens and home (residential) food gardens. Commercial operations are mostly farms, but these can vary in size and specialty. The goals and objectives of the food growing operation will delineate the types of inputs and outputs that are important to it. Before proceeding further, it is necessary to clarify some definitions with respect to farm size, cost and expenses.

Food Growing Operation Size

The term "small farm" has changed through time, but the most notable definitions applied in the United States include:

- The United States Food and Agriculture Act of 1977 defined a small farm as one that has less than $20,000 annual sales (USDA 1978)
- The United States Department of Agriculture defined a small farm as a retirement or residential farm that has less than $250,000 annual gross receipts (USDA 2010); this is the definition that is in common use today

It is important to note that these definitions lack any reference to the spatial size of the farming operation and refers only to economic status. However, in some cases, the farm's size is a crucial consideration when evaluating expenses and profit. For example, a two-acre farm may produce $25,000 worth of gross receipts annually because it is an efficient operation that produces a crop that demands a high price. In contrast, a soybean farmer may need as much as 40 acres to generate the same amount of gross receipts (University of Illinois at Urbana-Champaign 2013). Both of these operations are defined as small farms even though they differ significantly with respect to the amount of gross income generated per unit area, expenses (labor, materials and other investments), growing methods and the amount of land required to provide sufficient income. Although this example was taken from a rural agriculture operation, it does illustrate how size and profitability are not necessarily correlated, except when comparing operations that produce the same crop type using the same growing methods. This principle also applies to urban agriculture.

The size of commercial farms, non-profit farms and subsistence farms (in urban and near-urban areas) can range from one acre (small-scale farm) to dozens of acres or more. Because of relatively high land prices and development pressure, urban and near-urban farms that are a few acres (or more) in extent are much less common than small-scale farms. In contrast, community gardens and residential food gardens can range from a few acres (community gardens) to less than 50 square feet in extent; these smallest of food growing operations will be referred to as "very small-scale" operations.

Costs and Expenses

The term "cost" can have different meanings based on context and use. In business, it often refers to the amount of money that was spent on a specific resource or need. However, the term has also been used in a broader sense to include things such as the impacts from pollution created by an activity. In subsequent chapters, there will be more about the business context and broader usages of the term "cost". To avoid confusion, throughout the remainder of this book the term "cost" will be defined in the broader sense- it is referring to the things that have negative financial and non-financial effects (harm), or are resources (esp. non-renewable) that are consumed in the process of conducting an activity. For the strictly monetary definition of cost, the term "expense" will be used to identify cash outlay or expenditures.

There are some expenses that are common to many food growing operations and some that are specific to a particular practice. When looking at the full range of food growing practices in urban and near urban areas, few expenses are universal and essential to all operations. For example, a commercial farmer who must generate a minimal profit each season is concerned with the price of seeds, fertilizers and labor, whereas a residential food gardener who uses organic methods, recycles organic wastes for fertilizers (compost) and save seeds for planting the next season is not accruing these expenses.

The costs involved with running a food producing operation are many and some can be difficult to quantify. Some of the costs include:

- Start-up expenses, which is the money spent to start the food-growing effort; this is always a primary consideration for commercial operations
- Operating costs, which is the money and effort spent to sustain the food-growing operation through time
- Environmental costs
  o Consumption of non-renewable resources
  o Environmental degradation or destruction
  o Ecological impacts
- Other costs:
  o Labor
  o Acquiring materials
  o Repairs
  o Emotional stress (e.g., exasperation when a crop fails)

When initiating a new food-growing operation, there will be expenses that will occur only during the startup period. These can include the purchases of long-lived equipment and supplies, such as a rake or filing cabinet to store important paper documents. But, once these purchases are made, there is usually a list of seasonal, on-going expenses that maintain the operation from season to season. A list of the typical on-going expenses realized by various urban and near urban agricultural operations is shown in Table 6.

Table 6. On-going expenses for urban agricultural operations.

Expenses by Activity	Medium- to Small-Scale				Very Small-Scale
	Commercial	Non-Commercial			
	For-Profit Farm	Not-For-Profit Farm	Subsistence Farm	Community Garden	Residential Food Garden
Season Start:					
Planting	Y	Y/N	Y/N	N	N
Fertilizing	Y/N	Y/N	Y/N	Y/N	Y/N
Maintenance:					
Fertilizing	Y/N	Y/N	Y/N	Y/N	Y/N
Irrigation	Y/N	Y/N	Y/N	Y/N	Y/N
Pest Control	Y/N	Y/N	Y/N	Y/N	Y/N
Materials/Tools	Y	Y	Y	Y	Y
Harvest:					
Crop Harvest	Y	Y/N	Y/N	N	N
Land Clearing/Prep	Y	Y/N	Y/N	N	N
Packing/shipping	Y	Y/N	NA	NA	NA
Income Stream:					
Marketing/Sales	Y	Y	N	N	N
Grant/Donations	N	Y	N	Y	N
Business:					
Taxes/Rent	Y	Y/N	Y/N	Y/N	N
Permits/Licenses	Y	Y/N	N	Y/N	N

Y= Yes, each season there is usually an expense for this activity
N= No, there is usually no expense for this activity
Y/N= There may not be expenses each season for this activity
NA= This activity does not take place for this type of agriculture

Commercial Farms

Numerous volumes have been written about farm economics and this book does not intend to add to that library shelf. However, a review of relevant economic factors is warranted so that a meaningful comparison between the forms of urban and near-urban agriculture can be presented. Commercial farms operate entirely (or almost entirely) within the realm of the commercial market system where profitability is influenced by two primary factors: production costs and the price that can be charged for products (which is influenced by product supply and demand). If the amount of expenditures exceeds that which is taken in through crop sales (the primary income source) or other income, then the operation is not profitable and is not economically sustainable through time. Although these factors may also exist in agricultural systems outside of the commercial economic setting, they are of reduced influence or are not important enough to be included in non-commercial economics.

Commercial farms exist to grow food for others to consume and in exchange for this service they generate income. Commercial operations usually have a sole source of income- from the sale of the harvested farm crops. Sales transactions can be made to the general public through farm stands, to wholesalers (produce suppliers, brokers) or sold directly to customers (restaurants, markets, institutions). The price that the produce is sold for is a function of demand, availability and additional costs that are added onto the farm price (such as brokers' fees). Reduced production or crop failure will impact the farm's income, as will growing conditions in other regions where competition from other growers may influence the price of certain crops. Because of these variables, the gross farm income from the sale of crops can change from year to year, making commercial farming a somewhat risky financial undertaking.

As compared with other forms of urban agriculture, commercial farms typically have a high number of activities that require cash outlay to be completed (Table 6). This is because they are either large operations (spanning many acres) and/or use high-intensity growing methods that require mechanical planting or labor that come with costs. Planting, fertilizing, soil preparation (e.g. tilling), harvest, packaging and product sales usually require mechanical equipment, hired labor and/or the purchasing of materials. For operations that use conventional growing methods, there may be additional expenses associated with the purchase and application of fertilizers and pesticides. The expense involved with some of these activities can be broadly categorized as fixed or variable.

Fixed costs are those that must be incurred by the farmer whether or not production takes place, such as financial records management, payments for land purchases or rent, and depreciation on farm machinery, buildings, and equipment (Debertin 2012). The cost of using a piece of land is the same if 1 bushel or 1000 bushels of food are grown on it because land taxes or rent must still be paid. *Variable inputs* or costs are those that do change because of fluctuations in an external or internal factor. An example of an *external factor* is one that lies beyond the control of the food grower or farmer. One such example is rainfall, which can affect the amount of supplemental irrigation (which can be expensive) that is needed. An example of an *internal factor* is fertilizer- as more food crops are planted on a piece of land, more fertilizer is needed to maintain productivity. These internal factors are influenced by actions the farmer takes; for example, planting more crops necessitates the purchase and application of more fertilizer.

The price of *variable inputs* can change according to the amount purchased. Bulk purchases of supplies and materials (e.g., seed, fertilizer, herbicides, and insecticides) can reduce the per-unit price. In other cases, such as irrigation from self-supply wells, the main expense is for the initial purchase and installation of pumps, pipes, irrigation heads and control devices. After this initial investment, the differences in cost between 1 inch, 3 inches or 5 inches of applied irrigation water is usually low.

With regard to some variable inputs, there is a phenomenon that is referred to as the law of diminishing returns. For example, a farmer can increase the crop (e.g., corn) yield by applying a moderate amount of fertilizer (e.g., nitrogen) to the field. However, there comes a point that with each incremental increase in fertilizer additions, there will be progressively less increase in crop yield. In fact, a threshold may be reached where the level of fertilizer in the soil becomes toxic or causes micronutrient-limiting conditions that can reduce crop productivity. Were it not for the law of diminishing returns, a single farmer could produce all the corn required to feed the world, merely by acquiring all of the available nitrogen fertilizer and applying it to his or her farm (Debertin 2012).

Some fixed and variable costs are difficult to define. The classification of a specific input as a fixed or variable cost item is related to the particular time period that is referenced. This is because over a long period of time, items that are considered fixed costs need to be replaced or changed. Over the lifetime of a farming operation, the farmer is able to trade, buy and sell land, machinery, and other inputs into the production process that would generally be considered fixed. Thus, over very long periods, all costs are normally treated as variable (Debertin 2012).

One type of cost that food-growing operations rely on are resources. Some well known resources include energy, water and minerals but these are only a few of the resources needed to grow food. A broader definition of a resource, and one which is used in this book, includes monetary capital available for investment, materials (natural and man-made), labor, skills that the laborers possess, and other things that are utilized or consumed during the food growing activity. Some resources are consumed during the food-growing activity (e.g., irrigation water) and some are renewable (e.g., labor). Resources come with constraints and limitations on availability (Debertin 2012). The acquisition and use of a resource usually comes with expense.

One type of commercial farm operation that has become popular is Community Supported Agriculture (CSA). The CSA defers the growing season operating costs to customers at the beginning of the growing season. In essence the CSA customers subsidize the farmer, who then provides weekly shares of the harvest to customers. This approach to commercial agriculture reduces the potential for financial losses to the farmer as customers assume the risk of crop failure.

Non-Commercial Food Growing Operations

Non-commercial food growing operations are very different from commercial operations and, as such, cannot be described or evaluated on the same terms. As previously discussed, urban non-commercial agriculture is influenced by environmental and ecological factors that are distinct from rural commercial operations. Some of the other differences between commercial and non-commercial food growing are:

- Why it is practiced
- Who conducts it
- How it is practiced
- The conditions under which it is practiced
- The role it plays in society

One striking difference between non-commercial and commercial operations is that the latter exists to generate a profit and the former exists to provide goods and services to the gardener or local community. The output from commercial operations is sold to generate income, whereas the output from non-commercial operations is used to avoid spending money on food. Commercial operations run on the profit generated from previous growing seasons and if no profit is realized over a period of time, the operation will cease. Non-commercial food production reduces dependency on products that are supplied through a commercial market system and, as such, can contribute to individual and community economic sustainability. More about the economic structure of three types of non-commercial agricultural operations (non-profit urban farm, community garden and residential food garden) are discussed below.

Farms

Non-profit farms within and adjacent to urban areas have been established for a variety of reasons and, for the most part, are as individual as the entities that run them. Some are long-held family-owned subsistence farms that have been slowly encircled by development. Others are non-profit farms that have been set up to meet a need in the community. Not all non-profit farms are entirely divorced from agricultural markets. Goods and services may be traded for excess farm products through bartering or sold locally through informal transactions. More about subsistence and not-for-profit urban farms are provided below.

Subsistence Farms

Subsistence farms are operated to provide sufficient food for the farmers and their families. Investment in the farming operation is mostly limited to maintenance of the existing infrastructure and annual purchases of seeds or supplies that may be needed to produce the next season's crop. As compared to commercial farming, subsistence farming is less expensive to operate because it does not seek to generate a profit. The economic workings of subsistence farms may resemble that of a commercial farm since the owners rely on the operation to provide for their family and there is a vested interest in tracking costs. However, the necessity to produce a profit from the sale of goods is absent and market forces can have little to no influence on the operation's success.

From an economic perspective, the value of subsistence farming lies in what the family *does not* have to purchase instead of what has been earned through crop sales. This aspect of subsistence farming retains money within the household that would otherwise have been spent on basic necessities and reduces the minimum level of income required by the family. This aspect also distances the subsistence farmer from a commercial market economy that is based on cash, sales and income.

The number of expenses associated with subsistence farming is lower than for commercial agricultural operations (Table 6). Because subsistence farms typically support the families that work them, labor expenses are usually low or non-existent. There is no need for marketing, sales, packing, distribution or commercial licensing, which saves money and lowers operating costs. Additionally, if organic methods are used, the expense of purchasing fertilizers and pest control chemicals is reduced or eliminated. The reduced cost of subsistence farming makes it a financially efficient form of agriculture.

Crop selection, harvest and post-harvest activities conducted by subsistence farmers are heavily influenced by the family's heritage, culture and experiences. Subsistence farming gives greater freedom of choice with respect to the varieties of food that are grown as compared to commercial farming. Surplus may be traded or sold within the community. Rather than relying on continual purchasing of food products, food preservation techniques, such as home canning, drying and root cellaring (in appropriate climates) extend the length of time that farm products are available for consumption. Because of this, there may be less waste produced by the subsistence farm than through the commercial food market system, where losses can occur along the supply chain that connects the farmer, broker, market and consumer.

Non-Profit Farms

Non-profit farms exist for a number of reasons, but unlike subsistence farms, they are not focused on feeding the farmer's family. Non-profit farms include those that have been established to provide opportunities in underserved communities (e.g., employment, access to fresh produce), have been created as learning centers and those that have been established to stimulate activity on an underutilized parcel of land. Unlike subsistence farms, food production may not be the paramount purpose of the non-profit farm. Because of this, the financial benefit to the owners (reduced living expenses associated with essential food purchases) is not necessarily an important concern to the farm. Expenses vary from operation to operation, based on the farm's mission. For financial stability, many non-profit farms rely on grants and other sources of external financial support for funding.

Non-profit farms have been established in underserved communities and food deserts to improve living conditions in those areas. These farms often have a mission of making healthy and fresh fruits and vegetables available to local residents through community food programs (e.g., soup kitchens) or direct sales. These farms, if large enough, can also offer employment opportunities for local residents and become centers for community interaction. Some may operate or participate in farmers markets within the community. Some non-profit farms that sell produce accept state-level or federal-level food assistance program credits as a way to encourage the purchase of locally grown fresh produce by low-income families.

Some non-profit farms have been established as learning centers that teach food growing, food preservation, cooking and healthy living skills. As education-based facilities, they do not rely on the sale of crops for financial solvency. These education-based farms are often extensions of university departments, institutions, governing bodies or are created by a not-for-profit organization. Financially, they exist on donations, grant funding, income from use fees and overseeing institutional support. Besides education, these farms propagate a culture of self-sufficiency to improve the social conditions within the community.

Community Gardens

As with non-profit farms, community gardens have been founded for a variety of reasons and the benefits they provide to the community are in line with some kind of mission statement. A traditional form of community garden is one that has a number of individual plots that are leased to local residents. Community gardens offer space to grow food to city residents who do not have space of their own, opportunities for social interaction and access to the shared knowledge of other local food gardeners. The food grown within a single plot supplements the gardener's household food budget, but is usually not a major contributor to it. The gardener also benefits by knowing where the food came from and how it was grown.

Community gardens can be financially supported through a number of ways. In many cases, the land that community gardens are situated on has been offered for little or no cost by the landowner. Public support can be provided through land access, grants from charitable organizations and directed giving. Another way to support operations is by charging a participation fee, which sometimes includes a sliding scale that is based on family income level.

Residential Food Gardens

Home food gardens are typically created by, maintained by and benefit the household who owns the land it occupies. Residential food gardens have been around perhaps as long as humans have been living in cities and usually reflect the heritage, culture and food preferences of the household (Carmichael 2011). Like subsistence farming, the household usually consumes the food produced from residential gardens, but excess may be traded or sold. Residential food gardens benefit the household by decreasing the amount of food that is required to be purchased to sustain the family and lowering the minimum level of income required for the family to meet essential needs.

Residential food gardens are perhaps the most personal of all food growing spaces and can take on a myriad of forms and styles, reflecting the heritage of the grower and his or her culinary preferences. Beside crops plants, some residential food gardens also contain decorative, utilitarian and medicinal plants. These gardens can provide resources for the home, but may also have social and personal significance that goes well beyond the provision of food. Home gardens are often gathering spaces for friends and family, and are places of respite, places to go for solace or for observing nature. The ability to create or do things for one's self, instead of paying someone else to do a task, provides a sense of self-sufficiency that cannot be obtained from market products.

Other Food Growing Operations

A relatively small number of niche gardens and farming operations exist in metropolitan areas that do not fall into the categories described above. These exist as educational facilities, display gardens, and demonstration gardens, and to grow food to supply a specific user (e.g., restaurant, soup kitchen). These may be found on the grounds of educational institutions (e.g., grade schools, colleges), in public parks, botanic gardens, adjacent to restaurants or associated with social service-oriented organizations. For the most part, these food-growing operations exist for educational purposes or to directly meet the food supply needs of a single social or commercial operation.

Chapter 7. Quantifying Agricultural Inputs

The practice of cultivating plants requires a number of basic resources: space to grow, soil or growing medium, water, light, appropriate climatic conditions (natural or artificial), nutrients and care; these are collectively known as inputs. Different forms of agriculture approach the acquisition and management of these resources in remarkably diverse ways and each comes with unique costs. Quantification of the costs associated with the use of these resources is necessary to have a financially successful commercial food growing operation, but there is far less emphasis on developing markets and measuring costs by non-commercial operations (Table 5). This chapter will begin by identifying and describing some of the tangible costs associated with food growing operations. The latter part of the chapter will present ways that some of these costs may be quantified.

Food growing costs can be broken down into five essential activities: preparation, planting, growth, care and harvest. During this process, the cultivated organisms are provided a suitable growing environment, nutrition, protection from damaging pests and care. Each of these five activities requires the input of energy and/or materials and can have negative impacts (costs) beyond the growing site (Figure 5). Furthermore, some of these activities are supported by infrastructure, materials and other inputs that also come with costs.

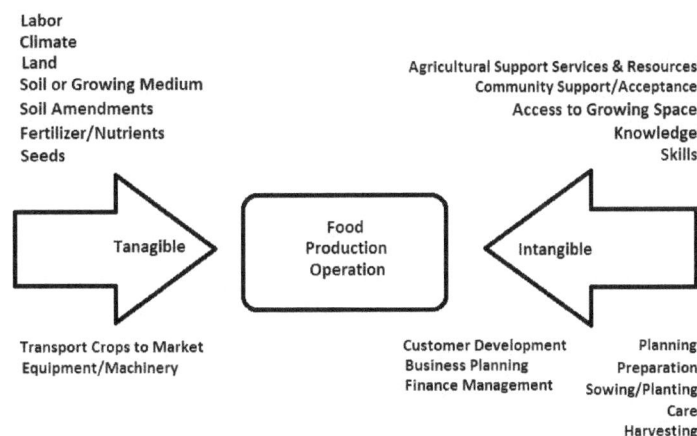

Labor
Climate
Land
Soil or Growing Medium
Soil Amendments
Fertilizer/Nutrients
Seeds

Agricultural Support Services & Resources
Community Support/Acceptance
Access to Growing Space
Knowledge
Skills

Tanagible

Food Production Operation

Intangible

Transport Crops to Market
Equipment/Machinery

Customer Development
Business Planning
Finance Management

Planning
Preparation
Sowing/Planting
Care
Harvesting

Figure 5. Inputs required by urban food growing activities; Some activities are exclusive to commercial operations.

Besides these essential activities, there are also non-essential inputs into food growing operations that do not contribute to the growing of crops but add other, often intangible, benefits or values. These costs may be things like labor expended to maintain a neat appearance, the installation of decorative elements to make a growing site more attractive or expenses associated with supporting activities and causes that cultivate relationships that form beneficial social connections. Although non-essential, these costs may be important for developing the business, customer relations and sharing information among peers.

Inputs can be divided into two categories: tangible and intangible. Tangible inputs include: capital (money), labor (time, people), land, soils, water, access, climate and equipment. These are highly measurable inputs and are frequently considered essential costs of commercial operations. Intangible inputs include: knowledge, skills, community support and agricultural support resources. Although these are identifiable, some are quite difficult to measure and it may be impossible to accurately quantify others.

Tangible Inputs to Food Growing Operation

Typical arable farming and forage crop production processes include tillage, fertilization, sowing, plant protection, irrigation and harvesting (Nemecek *et al.* 2007). Commercial operations may also include business activities (customer development, business planning, bookkeeping, etc.) and transport of crops to markets. Although the processes may be similar, the relative importance of any of these processes can be markedly different between different types of urban food production operations.

There are a number of tangible inputs associated with the practice of growing food. The most obvious are fertilizers and seeds. But, there are also other costs that may be less apparent and are no less important to be aware of. These can include planning, labor (e.g., time spent tending the crop) and control of diseases or pest outbreaks. These *visible costs* (those that can be seen by the grower) can be referred to as *direct costs* because they are directly related to the practice of food growing. There are, however, *indirect costs* associated with agriculture that are less obvious and may not occur near the site where the food is being grown. These indirect costs may also be referred to as *extended costs*- costs that extend beyond the realm of the farmer and farm field. Some indirect costs include impacts from pollution emanating from the growing site, environmental degradation caused by the food growing practices, ecological loss as production land is taken from natural habitats, and the loss of ecological goods and services associated with the loss of natural habitat.

As one embarks on the journey of food growing, there are a number of activities that must be undertaken before the first seeds can be planted. These include:

- Planning (How will the crops be grown? What will it cost?)
- Crop selection (What will be grown?)
- Preparation: Materials and infrastructure (What materials and infrastructure are needed?)
- Growing medium preparation

Planning

Planning is among the most important of activities. The time invested in planning is perhaps the most cost-cutting measure that can be made. Without proper planning, expenses can overwhelm profits, pitfalls will not be avoided and inefficiency will be an unintended consequence. Preparatory planning should include the following considerations:

- What are the growing site's conditions (e.g., exposure, soil type)
- What to plant (what can be grown at this site?)
- How much will be planted?
- What planting preparations are needed?
- Maintenance- labor and materials needed
- Harvest- how, when and what will happen to harvested produce?
- Calculation of financial limitations and goals (do I need to make a profit?)
- Identification of limitations (e.g., budget, maximum hours of available labor)

These considerations are fundamental to any growing operation and each comes with some level of cost. Each type of food growing operation will be concerned with these considerations in different ways and will think of them as having different levels of importance. Any consideration that is relatively unimportant to an operation can be disregarded. For example, concerns about how to, when to and what to do with harvest are usually negligible for a home food gardener with a small vegetable garden, but is a primary issue for commercial farmers.

Crop Selection

One of the most important decisions that must be made by the growing operation is what to plant. There are two approaches to this question and they come from vastly different perspectives. The horticultural perspective looks at a plant that one wishes to grow and then sets about manipulating the environment to accommodate the plant's needs. An extreme example is the construction of greenhouses or conservatories in temperate climates that allow year-round growth of tropical plants that would die with the first cold weather event. The second perspective comes from the ecologist, who first examines the local environmental conditions (e.g., climate, soils, local pests and diseases), then selects plants that are both adapted to those conditions and that will not be disruptive to local ecosystems. Most food growers tend to operate between the two perspectives by doing a measured amount of site preparation to accommodate a suite of plants that are fairly well adapted to their local climate.

Once the growing site's conditions have been assessed and the amount of investment in site manipulation has been ascertained, a menu of potential crop plants can be created. Some varieties of crops come with higher growing costs than others. Here are some general relationships between the crop plant type and cost:

- The plants that could be grown with the least cost (low-maintenance) are those that are adapted to the ambient conditions at the site and that have some degree of resistance to local pests and plant diseases. The more that the selected crop plants' needs diverge from the growing site's natural conditions, the more costly it will be to grow those plants.
- The expense involved in bulbs and transplants is greater than planting seeds.
- Short-season crop plants may be more cost-efficient to grow if a second crop can be sequentially planted.
- For commercial operations, there is financial incentive in growing plants that bring the most profit on the open market.
- Space-efficient plants may be more cost efficient than space-needy plants (e.g., lettuce vs. squash).
- Continual harvest varieties usually require a greater investment in labor than single-harvest crop varieties.
- Some crops are easier to harvest than others, which influences farming costs.

Crop selection is often revised throughout the life of a growing operation. Over time, experience with production and operation will inevitably lead to adjustments in what is being grown. This process of learning from experience and trying new methods to avoid past problems is referred to as *adaptive management*.

Preparation: Materials & Infrastructure

Once the crop types and varieties have been determined, then the operational support infrastructure, including growing site preparation, needs to be established before planting can begin. Infrastructure can include both tangible and intangible forms. For commercial operations, initial infrastructure can include a business plan, the acquisition of essential knowledge, appointment of a management structure, installation of irrigation systems, setup of maintenance tools and materials, site/soil, etc. For residential food gardeners, the only infrastructure may be raised beds and soil.

The costs of pre-planting preparations can be high for commercial or larger-scale operations as compared to community and residential gardens. This is because the physical infrastructure needed to support larger-scale operations is typically more extensive and requires periodic repair or replacement. Following are some examples of expenses that may be accrued through time:

- Construction Phase: Construction materials (e.g., wood, pipe, sheet metal), labor, transportation of materials, etc.
- Utilization Phase: Infrastructure maintenance, energy, water (cleaning, drinking), miscellaneous materials (cleaning supplies, lubricants), etc.
- Waste disposal (during initial construction, repair or replacement): Dismantling, transport of waste materials, recycling, etc.

Besides infrastructure, some agricultural operations require machinery or tools to assist with planting, maintenance or harvest. Some of these can include:

- Tillers (plough, harrow, roller)
- Tools (pruners, saws, hoes, rakes, shovels, etc.)
- Tractors
- Harvesters (combine, harvester)
- Trailers and transporters
- Tankers (vacuum tanker, pump tanker, sprayers)
- Misc. agricultural machinery (seeder, mower)

Of course, costs will vary by the size of the operation (larger operations will require more expensive agricultural machinery), the growing methods used (e.g., traditional or industrial) and the crop plants that will be grown. Conceptually, the amount of site preparation someone is willing to undertake is usually related to the amount of available resources (labor, materials, funds, etc.). Once a resource becomes limiting, the scale of the plantings will either be reduced to come in line with the area that can be prepared or the menu of crop plants will be adjusted towards varieties that require less up-front investment.

Before planting can commence, there must be a sufficiently large area to accommodate the desired number of crop plants and the growing medium must be readied to accept the plantings. Almost all of the fruits and vegetables that are consumed by humans have been grown in soil or in a soil-like medium. However, an increasingly large amount of food is being produced using alternative methods that use a sterile medium that is not soil; one such example is hydroponics.

Soil Preparation

The soil condition at the growing site should be assessed to determine what actions should be taken before planting. The physical and chemical characteristics of the soil need to be tested to determine what kind of conditioning may be necessary to ensure adequate plant growth. Some soils, such as clay, are dense and can impede root growth; these soils need amendments to better support crop plant growth. Other soils, like sands, are nutrient poor. Other soil amendments can be added to improve water holding capacity, drainage, organic matter content and pH status. Besides soil condition, weeds and obstacles (such as large rocks) need to be cleared from the growing site.

Soil-Like Medium Preparation

Some food producing operations grow food in containers, such as buckets, trays, jars (sprouts) and pots. The preferred growing medium for these operations is a sterile soil-like medium that has been prepared from natural and artificial materials. The benefits of these mediums are that they are free of pests, they have been infused with nutrients, the medium allows for good root growth and the soil moisture holding capacity reduces the chance for drought stress. Purchasing these growing mediums can be expensive, particularly when a large quantity is needed.

Non-Soil System Preparations

Hydroponic systems, which were described in an earlier chapter, require setup and testing to ensure that the various pumps and tubing is functioning properly. These setups can come with a relatively high cost and a high yield is needed to offset the system's expense.

Planting

The planting of crops involves the purchase and installation of propagules (seeds, cuttings, seedlings, bulbs, etc.), labor, water and may also include the addition of growth-promoting agents (fertilizer, antimicrobial, mycorrhizal symbiont, etc.). The investment in these items depends on the crop types and the size of the operation. Typically, larger-scale operations use machines and equipment to assist in crop planting but smaller-scale operations may plant by hand.

Some agricultural operations till the soil prior to planting to increase aeration, reduce weeds and to create furrows into which propagules are planted. Other operations are "no-till", meaning that the soils structure is left intact and propagules are planted as plugs or drills that place seeds directly into the soil. Agricultural machinery is used for both of these planting methods. Hand planting is necessary or desirable with some crop varieties, particularly transplants or slips. This is obviously a labor-intensive and potentially cost-prohibitive activity for larger operations but it is a standard procedure for home gardens, community gardens and other forms of civic agriculture.

Growth

In order to produce harvestable crops, plants require certain conditions to thrive. Besides the growing medium, the main resources required for adequate plant growth are water, solar radiation and nutrients (Nemecek *et al.* 2007). Most agricultural operations seek to maximize productivity by providing these three things in the quantity and quality that stimulate optimal plant growth.

Water

In some areas, rainfall is usually not a limiting factor for agricultural production. Where it is limiting, irrigation is required. When supplemental water is required for adequate plant growth, this can come with a significant expense. Irrigation infrastructure can range from a hose and nozzle (small operations) to a complex system of pumps, pipes, spray heads, sensors and timers (larger operations).

Besides water quantity, water quality is an important consideration for agricultural operations. In drier climates, evaporation rates from soils and plant leaf surfaces are relatively high. Trace minerals or salts in irrigation water can build up and alter soil chemistry. In some cases, these minerals or salts can reach toxic levels and impact crop productivity. Water quality testing and soil chemistry monitoring may be necessary in climates that rely on irrigation, rather than rainfall, as a primarily water source.

Light

For proper growth, plants require an adequate duration and quality of light. Sunlight is free, but too much can burn crops and cause excessive evapotranspiration. Artificial lighting must provide the proper wavelengths that support photosynthesis. Outdoor farming operations in warm climates sometimes erect shade cloth canopies to reduce sun exposure and wilting during mid-day exposure. Indoor operations may need to provide an artificial light source so that adequate plant growth can occur. These indoor systems can come with a significant cost in terms of fixtures, lamps and electricity.

Nutrients

Throughout the growing season, plants extract nutrients and dissolved minerals from the soil or growing medium, which must be replenished from time to time to maintain productivity. Fertilizers are added to increase soil fertility and may be applied to increase the concentrations of a few specific nutrients or, as in the case of broad-spectrum fertilizers, a comprehensive suite of plant nutrients. Some of these include:

- The primary nutrients of nitrogen, phosphorus and potassium
- Secondary nutrients (sulfur, magnesium and calcium)
- Micronutrients (iron, manganese, boron, chlorine, zinc, copper and molybdenum)

Purchasing and applying nutrients comes with expense that is proportional to the area being farmed and the nutrient uptake rate of crop plants.

Care

From the first planting until harvest, growing plants requires constant oversight and surveillance. When pests are detected or damaging weather is predicted, the grower must mobilize protective measures to reduce crop damage. This constant oversight requires attentiveness, responsiveness and preparedness that are usually exercised through labor costs. When pests are detected or protection from threatening inclement weather is necessary, expenses are usually realized in the form of pesticides and protection devices.

Monitoring

Farmers must continually monitor plant growth and maturation throughout the season, looking for indicators of stress and pests. After a severe weather event passes, the farmer must survey the crop plants for signs of damage. These activities are usually carried out by direct observation and labor is the greatest cost.

Pest Control

Pest control is a part of most food growing operations and can be roughly divided into different types. Some pest control measures are preventive. These seek to eliminate or reduce the chance of an outbreak. Other pest control measures are remediative. These measures seek to eliminate the pest after it has been found in the growing area or to control pest numbers below a threshold where economic injury would occur (the *economic injury level*, or EIL). The control of pests below the EIL maintains crop productivity without the cost of complete pest eradication. This approach also keeps labor and materials costs to a minimum while maintaining economic benefit.

Pest control measures will vary according to the type of operation. Organic and traditional agricultural practices rely on integrated pest management (IPM) activities that include crop rotation, natural biological controls and sanitary practices to reduce the risk of pest outbreaks. All of these activities are carried out by labor that is above and beyond the basic effort needed to plant and harvest crops.

Pest control measures used by non-organic operations and especially industrialized methods usually involve the use of chemical pesticides and products that must be purchased. A pesticide is defined as "any substance or mixture of substances intended for preventing, destroying, repelling or mitigating any pest" (United States Environmental Protection Agency 1947). Pesticides are broadly categorized according to the target organisms: herbicides (weed control), insecticides (insect control), fungicides (fungal pathogen control) and others (nematicides, bactericides, rodenticides). Application of these products may be done with labor and machinery, depending on the size of the growing area.

Protective Measures

Inclement weather occurs from time to time and extremely strong winds, drought, excessive rainfall, damaging frost and wilting heat are normal parts of most climates. Protective measures can include shielding, temperature moderation, irrigation, flood mitigation and other activities. All of these come with labor and material costs that can be significant.

Harvest

Industrialized, large-scale farms that produce corn, soybeans, wheat or oats are usually planted and harvested using machinery. Urban agriculture usually grows a wider range of crops than these industrialized farms and relies more heavily on labor to harvest. Other than labor costs, there are expenses related to containers, machinery (such as trucks) and materials.

Intangible Costs of Food Growing Operations

The tangible costs of growing food are highly visible and relatively easy to quantify. Conversely, the intangible costs are often overlooked even though they can exert a considerable toll on farmers. Some intangible costs include:

- The acceptance of risk. There is an inherent risk in growing food because of the many uncertainties that can influence production. The risk is more pressing when a minimum profit must be realized to keep the operation financially viable.
- The burden of responsibility. Growing food requires continual oversight, surveillance, management and reactive responses to minimize problems that can reduce productivity. Although lower levels of responsibility may be enjoyable, high levels of responsibility can be tiring and can take the joy out of growing food.

- Food growers sometimes report feeling disappointment and frustration, particularly when they experience crop failure or uncontrollable pest outbreaks (Zahina-Ramos 2013).

Should the intangible costs of any food growing activity exceed that which the grower is willing to endure, they will likely abandon the activity. Identifying intangible costs is often not easy; placing a quantitative value of those costs is even more difficult. Intangible costs are often far more difficult to place monetary values on than tangible costs. Because indirect and intangible costs are not as easily captured as tangible costs, their valuation relies on a set of assumptions that may vary from place to place. Even still, it is worthwhile to place them within a valuation framework so that the magnitude of their impacts may be understood.

Quantifying Food Production Inputs

In order to estimate the profitability of a food growing operation or the cost-effectiveness of growing method, it is necessary to quantify and summarize inputs. The magnitude and cost of agricultural inputs vary according to the type of operation and some inputs are difficult to place a market price on. Although there are challenges to placing quantitative values on intangible costs and benefits, a number of methods have been devised to address this issue and some deserve consideration.

The various steps involved with food growing operations were described in the previous sections, along with possible costs associated with these activities. These costs may be aggregated and sorted into related types of inputs. Furthermore, these costs can be divided into those that are one-time startup costs and those that are continual throughout a season. By organizing costs in this way, they can be summed and their relative contribution to the food growing operation can be quantified.

Direct and Indirect Costs

Costs may be divided into two types: direct and indirect. Direct costs are those that are explicitly input to the operation. These include labor, materials, water, fertilizers and monetary investment. These are usually easy to track and to place values on.

Indirect costs are often overlooked (or unknown), but can be significant. Indirect costs may have values and expenses that lie beyond the monetary system. For example, chemical phosphorus fertilizer is relatively inexpensive in terms of monetary cost. But, the environmental costs may be substantial and difficult to capture in a traditional pricing scheme. Other indirect costs involve the CO_2 emissions resulting from the use of fossil fuels, environmental degradation that resulted from the mining and processing of a phosphate fertilizer or the impact of non-native species that escape cultivation. The methods used to place values on these indirect costs are not addressed in this chapter, but are presented in later chapters that deal with cost-benefit analysis and the residential food garden case study.

Materials and Supplies

Expenditures from purchased materials and supplies can be obtained from receipts and financial records. In cases where these records have been lost or have not been kept, the list of products that were purchased can be valued by referencing prices for these products at local retail outlets. For estimates of expenses expected for a future growing season or from a proposed food production scenario, a list of required materials can be created and the cost of those products enumerated by referencing prices for these products at local retail outlets. These prices can be adjusted to account for expected or estimated variations in the value of currency if projections beyond one or two years are being done.

Labor

The total amount of time spent by all participants working on the food growing operation throughout all phases is collectively part of labor costs. There is a set of activities that must be carried out when the operation is being initially established and this labor is usually considered as part of startup costs. Labor investment at startup can include planning, development of a business plan, establishment of a base of operations, materials purchase, infrastructure installation, client development, site plan approvals, etc. A breakdown of time/labor investments is shown in Table 7. There is another set of activities that are carried out seasonally or are on-going. A breakdown of these ongoing activities is shown in Translating labor costs into salary expenditures (or labor value) requires that the analysis knows how much time has been expended and how that time is to be valued. In addition, not all time is valued equally; for example, a manager's salary is typically much higher than the salary of a seasonal worker. Some labor expenses (for commercial farms) can be calculated from pay stub records or inferred from an employee's pay rate per hour. Care must be taken to include not just salary, but the total compensation extended to the employee (including sick leave, vacation leave, holiday pay, insurances and contributions to retirement funds), which can be from twice to three times the salary.

The per-hour value of someone working a home food garden or a community garden plot is more complicated to ascertain. One approach is to survey the gardener and ask them what they perceive his or her time to be worth. This worth can be based on how much money they would have made if they were working their regular job, the money savings realized by gardening or how much they would have to spend to hire someone to do the gardening for them.

Table 7. Time/labor investment parameters.

Activity	Parameters
Planning	Sizing the next season's operation, determining planting schedule, mapping out where to plant and other related issues.
Garden or farm maintenance	Preparation of soil for planting, transplanting seedlings, sowing seeds, mulching, weeding, applying soil amendments (fertilizer, mulch, manure, compost), applying pest controls, and plant maintenance.
Irrigation	Time spent irrigating. Time investment varied by watering method.
Solarizing	Time spent solarizing garden soil to reduce soil pests (warmer climates only).
Construction	Time spent building and repairing the physical components of the garden beds, such as construction of raised beds and drip irrigation installation.
Mulch/manure	Time spent obtaining mulch/manure (excludes application time).
Composting	Time spent composting on-site (excludes acquisition and application time).

Table 8. On going and seasonal inputs to food growing operations, by activity.

Activity	Inputs			
	Time (Labor)	Money (Capital)	Water	Other
Planning	S	N	-	Knowledge, Experience
Preparation Soil Soil-Like Medium Non-Soil Medium	S S S	S S S	N N N	
Planting	S	S	S	Knowledge, Experience
Growth Water/Irrigation* Fertilizers Lighting*	S S	N S S	S S -	
Care Monitoring Pest Control Protection	S S S	N S S	- N N	Knowledge, Experience, Observation
Harvest Sales Preparation* Marketing*	S S	S S	S -	Knowledge, Experience, Record-keeping
Business Tracking/Accounting* Taxes/permits/licenses*	S S	N N	- -	

S= significant costs
N= costs are relatively small
*These costs are not a part of all food growing operations

The time and labor investment parameters in Table 7 can be calculated from reasonable estimates or recorded labor data collected during a previous growing season to estimate expected labor costs, but only with caution. Historic data from only one or two seasons cannot characterize the full range of conditions that normally occur over many years. To reliably estimate the potential future annual labor costs, it is best to collect data over many years and calculate an average value (and range) for a parameter. One may also report expected labor costs with a margin of error that can estimate the upper and lower bounds of an expected range of values.

Another source of time and labor investment data could come from similar operations within the same region. When considering the use of data from other regions, the growing conditions may be significantly different and caution must be used before adopting those values for the analysis. Some agricultural universities, through their cooperative extension offices, publish data on the amount of labor required to grow a specific crop type (Table 9). The analyst should consult the local agricultural extension office or state agricultural university resources for labor tables based on data from their region. Because these data may originate from larger-scale commercial operations, they may not be appropriately applied to small-scale operations.

Table 9. Estimated average labor needed to grow selected crop types.

Crop	Hours/Acre[1]
Beans (string)	109.2
Broccoli	136
Cantaloupe	108.9
Corn (sweet)	71.7
Cucumber	153.2
Eggplant	207.8
Okra	301.8
Pepper (bell)	185
Potato	105.6
Raspberry	100.7
Spinach	103.4
Squash	140.7
Strawberry	140.3
Tomato	306.4
Watermelon	76.5

[1]Data source: University of Missouri Cooperative Extension, *et al.* 1965;

Translating labor costs into salary expenditures (or labor value) requires that the analysis knows how much time has been expended and how that time is to be valued. In addition, not all time is valued equally; for example, a manager's salary is typically much higher than the salary of a seasonal worker. Some labor expenses (for commercial farms) can be calculated from pay stub records or inferred from an employee's pay rate per hour. Care must be taken to include not just salary, but the total compensation extended to the employee (including sick leave, vacation leave, holiday pay, insurances and contributions to retirement funds), which can be from twice to three times the salary.

The per-hour value of someone working a home food garden or a community garden plot is more complicated to ascertain. One approach is to survey the gardener and ask them what they perceive his or her time to be worth. This worth can be based on how much money they would have made if they were working their regular job, the money savings realized by gardening or how much they would have to spend to hire someone to do the gardening for them.

Monetary Investment

Cash expenditures over a growing season can be recorded to yield a tabulation of actual costs (after the fact) for a particular operation. Examples of expenditures can include the purchase of seeds and propagules, fertilizers, pesticides, mulch, potting soil, pots, containers, wood for raised beds, irrigation supplies, signs, stakes/supports, etc. Table 10 lists the typical expenditures realized by urban food growing operations.

Table 10. Typical expenditures of urban food growing operations.

Investment	Description
Building materials*	Growing containers, lumber, hardware, irrigation supplies, etc.
Plant material*	Seeds & starter plants
Pest control*	Pesticides
Soil and soil amendments*	Potting soil, garden soil, fertilizer and soil amendments
Supplies*	Garden implements, tools, signs, stakes and plant supports, compost bin, etc.
Machinery & equipment*	Tillers, mechanical planters, heavy equipment, etc.
Aquaponic/hydroponic* systems	Trays, troughs, racks, distribution lines, etc.

*Natural resource, energy and manufacturing costs for these products vary, but they can be generally ranked from less to most costly. (I need a section somewhere that ranks these and describe differences, as well as the life-cycle assessment of each).

As with labor costs, measured data can be used to accurately calculate the actual expenditures for materials and supplies for a previous growing season. If similar conditions are anticipated to occur in a future time frame (such as the next growing season), these measured data can be a reasonable estimate of what to expect.

However, when estimates of materials and supplies are used, caution is warranted with respect to the reliability of these estimates under different environmental conditions. Understanding potential variability or range of expenditure values can help the analyst bracket possible costs.

Water Usage

Access to water is essential for aquaponics and hydroponic systems and supplemental irrigation for in-ground or container food production is necessary in some climates. The costs involved with acquiring and supplying water of sufficient quality and quantity varies from region to region, and by water source. Some water sources most often used by agricultural operations are shown in Table 11. Each of these comes with varying levels of cost. The most expensive water source is from municipal sources, but it is also among the most reliable.

Water use can be measured using the following means: (1) direct measurement using a meter or water measurements before application, (2) indirect measurement using a rate (pump output expressed as a volume per unit time) applied to a measured time of pumping. To ensure accuracy of measurements, the meter or pump output rate needs to be verified on a periodic basis using acceptable standards.

Table 11. Water sources typically available for urban food production.

Source*	Description
Municipal potable	Utility-supplied potable water from a regulated source that usually comes with a rated cost per unit usage plus taxes and fees.
Municipal non-potable	Utility-supplied non-potable water, such as treated wastewater, which is sometimes supplied at low or no cost.
Self-supply well	Water obtained through an on-site groundwater well. On-going costs include electrical pump operation and maintenance.
Self-supply surface water	Water obtained from a nearby surface water feature such as a reservoir, lake or canal. On-going costs include electrical pump operation and maintenance.
Self-supply rainwater	Rainfall runoff that has been captured and stored on-site. These systems typically have low or no operating costs.

*Natural resource, energy and manufacturing costs for these products vary, but they can be generally ranked from less to most costly. (I need a section somewhere that ranks these and describe differences, as well as the life cycle assessment of each).

Calculating (from historic data) or estimating water costs is simple and straightforward when one uses a municipal or rainwater source. One usually receives an itemized bill from the water provider that provides a good measure of actual water use. Operations that use rainwater usually rely on storage tanks that have a fixed capacity and use can be measured as it is stored or as it is applied. Self-supply wells and surface water usage can be metered or measured, but these sources usually require the use of electricity, infrastructure and, in some cases, permitting requirements that add costs that must be included in the consideration of cost.

Chapter 8. Quantifying Agricultural Returns

Those who grow food, whether they are commercial farmers or backyard food gardeners, do so because they receive some kind of valuable return (such as products or benefits) from the activity. Some of the benefits from agricultural operations include the production of goods (e.g., food, flowers, materials), the creation of jobs and contributions to the economy. There are many benefits from agriculture that lie beyond these tangibles. Many farmers farm because it carries on a family tradition that they deeply value. Some farmers feel fulfillment by growing food that feeds others. Community gardens are often great repositories of local food growing knowledge. Home food gardens reflect the heritage of the gardener and help to continue a food tradition. These benefits, which at times are also referred to as *returns* or *effects*, are usually taken for granted but, collectively, provide important positive influences to individuals and society. The term "returns" is usually associated with economics, especially when referring to the monetary return on an investment. The term "effects" is used in a broader sense and describes the effects that result from a particular action, some of which may be undesirable.

The returns realized from urban agriculture vary according to the type of operation. As with inputs (described in the previous chapter), these returns have been categorized into two distinct groups: tangible and intangible. Tangible returns are further divided into those that are directly realized by the farmer (i.e., internal to the food growing operation) and those that lie outside (or mostly outside) of the farming operation (Figure 6). Examples of internal tangible returns include the production of crops and the monetary capital that is generated by growing a crop. External tangible returns include contributions to the community's economy, environmental protection or ecological enhancement.

The most important internal tangible return from any agricultural operation is the harvested crop, which may be fruits, vegetables, herbs, flowers, medicinal plants, fibers and other plant products. Another important internal tangible return is the creation of monetary capital, which is the financial value of the crops. Commercial and non-commercial operations benefit monetarily from crops in different and important ways. In contrast to utilization of the physical crop, the conversion of this crop to currency or monetary value is a main objective of commercial agricultural operations. Non-commercial agricultural operations create monetary capital for the growers but are valued by the amount of money *not spent* to purchase food. For example, growing food increases a household's economic status because less money is spent on commercial plant products, leaving more money in the household budget to be spent on other things.

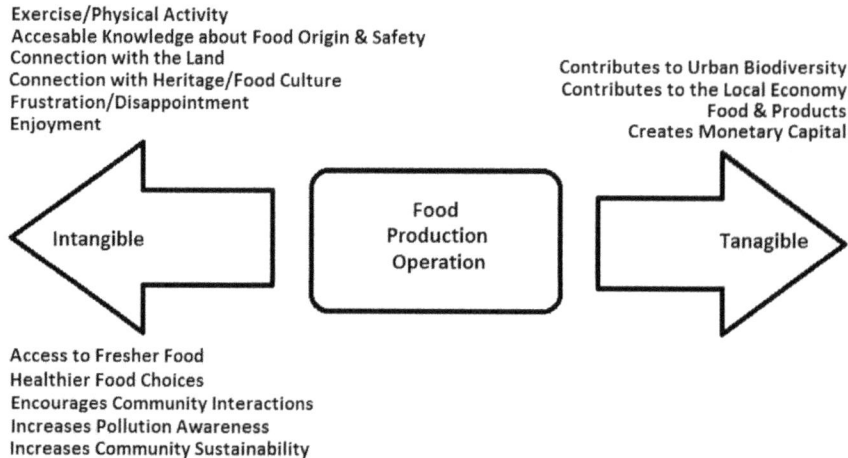

Exercise/Physical Activity
Accesable Knowledge about Food Origin & Safety
Connection with the Land
Connection with Heritage/Food Culture
Frustration/Disappointment
Enjoyment

Contributes to Urban Biodiversity
Contributes to the Local Economy
Food & Products
Creates Monetary Capital

Intangible

Food Production Operation

Tanagible

Access to Fresher Food
Healthier Food Choices
Encourages Community Interactions
Increases Pollution Awareness
Increases Community Sustainability

Figure 6. Tangible and intangible returns realized (i.e., effects caused) by urban food growing operations.

Measuring Tangible Returns

The internal tangible returns described above can be measured with relatively good precision when certain parameters are clearly defined. The physical plant products that are harvested from an agricultural operation are easily measured, once one determines *what* should be measured. This is not always a straightforward answer, but it must be defined before one can ascertain the financial value of the crops. The first step to measuring returns is to determine what the crop is.

Defining the Crop

In the agricultural sciences, productivity is usually meant to infer how much of a crop is harvested. What complicates this measure is the fact that the part of a crop plant that is normally used varies by region, as well as demographic, cultural and ethnic group. This makes the question of what is considered to be "food" not always easy answer. For example, one can easily determine the part of a cabbage plant that is food- it is universally accepted to be the dense head that forms at the apex of the short stem. But, what is considered to be the food part of a plant is not as easily defined in other vegetables. For a particular plant, one culture may traditionally eat only the fruit but another culture may use the fruit, flowers and leaves.

For commercial agriculture, a crop is usually defined as the plant product that has market value. The interplay between market system and commercial agriculture is an interesting and forever changing one. Commercial farms will plant crops that yield profit and that are reasonably easy to grow. When production is in excess of demand, prices paid for the crop decline because there is a surplus. When the market value of a commercial crop falls below a threshold where profitability is no longer possible, the farmer must resort to other measures to stay economically viable or change operations to include other more profitable crops if they want to remain financially solvent. Some of these more profitable crops are often specialty fruits and vegetables, which have smaller sales but higher profit margins. As farmers adopt these specialty crops and they become more abundant in the market place, the crop prices will fall along with profitability. However, the drop in specialty crop prices can also drive greater sales since consumers often welcome the addition of new and novel items into their diet.

For the non-commercial food grower, market prices and the availability of many commercial crops are of little importance. This grower's focus is centered on cultivating the types of produce they wish to consume or to pass along to others. These growers may also consume parts of the food plants that are typically not offered for commercial sale because they are not profitable. The definition of the crop may be broader for the non-commercial producer than for the commercial farmer. For example, in the United States today, the bulbous part of the kohlrabi plant is eaten but the large leaves are discarded. However, in the 1800s, the leaves were considered to be a desirable part of the plant and recipes for preparing kohlrabi leaves can be found in cookbooks from that era.

Non-commercial food growers may also produce plant varieties that are not offered for sale through markets. One example is the growing (intentional or unintentional) of wild or weedy edible plants. While some commercial operations seek to suppress the growth of these plants, growers that use organic or traditional methods may be tolerant of edible wild plants and may include them in the food harvest. Because non-commercial food growers may consumer plant parts or plant varieties that are not included in the commercial markets, non-commercial food growing operations may record higher productivity and food production potential than commercial operations.

Measuring and Expressing Productivity

Not everyone defines the term "agricultural productivity" the same way. To the farmer, it means the amount of marketable crop that a profit can be realized from. To the home food gardener, it means whatever comes from the garden that can be eaten. To the social scientist that is interested in relieving food insecurity, productivity is placed into a context of balanced and appropriate nutrition. Given these differences, each entity may define and quantify productivity in a different way.

Productivity is an important measure in all of the life sciences. It tells scientists if one set of conditions is better for an organism's or population's health or well-being. Productivity studies are used to determine how much of a crop is produced or if the application of a certain fertilizer increases crop yield. In ecological studies, the crop may be defined as all plant biomass. In agricultural studies, the crop is often defined as a specific part of the plant that has value to the grower. In both ecological and agricultural studies, gross plant productivity may be measured for an entire area or productivity measurements may be highly focused to determine the percent of a plant's biomass (or nutrient concentrations) has been allocated for roots, stems, leaves, flowers or fruit. In the latter case, this kind of information is helpful for plant breeders to determine which varieties invest more resources in the crop portion of the plant rather than in parts of the plant that have no commercial value.

Agricultural production has often been estimated from survey data collected from farmers across a large region. Major problems exist with this estimation approach. One problem arises from a lack of controlled experimental conditions that can sieve out effects of uneven distribution of factors such as weather conditions. In addition, other factors that may vary from farm to farm include soil conditions, experience/knowledge, management skills and investment in infrastructure. Most researchers recognize that the lack of controlled experimental conditions represents a major problem with this approach to estimating agricultural production functions and attempt to take steps to control for factors such as soil type and weather conditions (Debertin 2012).

Agricultural productivity is usually expressed as the amount of crop produced per unit area over time (e.g. $1kg/m^2/year$), which can easily be calculated from historic data. Productivity is important because it can also be used to infer how much may be expected from a given space when planning a farm or food growing study. The measurement of past agricultural production can be highly accurate for a specific site, but inferences beyond the study site can be problematic. The inference of future productivity is always imprecise, since many random and uncontrollable factors are at work.

Food growing operations often define a crop differently, which can lead to variations in productivity calculations for the same crop type. Commercial operations usually define a crop through the lens of commercial market commodities for a given food culture. In one culture, the tuber of the sweet potato is the crop of interest (what is sold for profit) and the leaves and stems of the plant are considered to be by-products. The amount of sweet potato tubers that can be produced from a finite space and the market value of those tubers form the basis for productivity measures for that commercial farm operation. In contrast, a commercial operation in another culture that also harvests the leaves and tubers of the sweet potato plant for commercial sale would measure productivity by including all of the edible parts of the food plant (leaves, stems and tubers). This operation would report productivity and profit values that are higher than the cultural area that only sells the tuber, even though the operations are growing the same crop plant. One must remember this difference when comparing productivity and cash crop values reported from different cultural regions. Likewise, the crop plant parts that are consumed may differ between non-commercial growers (subsistence farm, home food garden, etc.). Because of differences in food preferences across the world, those operations that use a broader definition of what constitutes a valuable plant product would report a higher crop yield and value from the same field of crops than operations that use a more restrictive definition.

The productivity of a farm field or food garden can be expressed in different units, depending on what is important to the investigator and what questions have been asked. To the commercial operation, it is important to know the amount of food that has been produced in terms of weight (mass) or volume. This is because commercial agricultural products are usually sold in these units. Volume is often an important consideration for shippers who need to know the amount of space a product will take up on a transport vessel. To the non-commercial food grower, weight and volume measures are seldom of concern except when one wants to document how much was donated to a cause or what might be needed to produce a secondary product (e.g., growing enough cabbage to make a batch of sauerkraut). Weight and volume are perhaps the most common metrics and they are easy to measure with conventional equipment. One needs only a scale whose accuracy is regularly verified using standard weights and a fixed volume measure.

To the nutritionist or social activist who wishes to understand how a community garden can improve nutrition in a food desert, the preferred metric may be the number of servings (cups) of certain types of food that were produced. Neither volume nor weight is very useful to determine how closely the agricultural operation's output can meet human dietary needs. For that application, productivity must be expressed in other units that are used by organizations that have explicitly defined how much food should be consumed for a healthy diet.

The U.S. Department of Health and Human Services and the U.S. Department of Agriculture jointly publish evidence-based dietary guidelines for Americans every 5 years. These recommendations have been developed to promote health, prevent chronic disease and to assist people with reaching and maintaining a healthy weight (Office of Disease Prevention and Health Promotion 2017). These recommendations form the basis of the United States Department of Agriculture's Choose My Plate guidelines (USDA 2016). These guidelines cover a range of food groups that are necessary for a healthy diet. The food groups included in the guidelines are: fruits, vegetables, grains, protein foods and dairy. Table 12 shows the weekly-recommended intake of fruit and vegetables outlined in the Choose My Plate guidelines; recommended amounts of grains, protein foods and dairy have not been included.

Table 12. USDA's Choose My Plate recommendations for fruit and vegetable consumption (cups per week).

	Fruit	Vegetables				
		Dark Green	Red & Orange	Beans & Peas	Starchy	Other
Children						
2-3 yrs.	7	0.5	2.5	0.5	2	1.5
4-8 yrs.	10.5	1	3	0.5	3.5	2.5
Girls						
9-13 yrs.	10.5	1.5	4	1	4	3.5
14-18 yrs.	10.5	1.5	5.5	1.5	5	4
Boys						
9-13 yrs.	10.5	1.5	5.5	1.5	5	4
14-18 yrs.	14	2	6	2	6	5
Woman						
19-30 yrs.	14	1.5	5.5	1.5	5	4
31-50 yrs.	10.5	1.5	5.5	1.5	5	4
51+ yrs.	10.5	1.5	4	1	4	3.5
Men						
19-30 yrs.	14	2	6	2	6	5
31-50 yrs.	14	2	6	2	6	5
51+ yrs.	14	1.5	5.5	1.5	5	4

Return Valuation

Besides the volume of produce harvested at the end of the season, another important measure is the value of the harvested crop. Before a crop can be valued, one must define what one means by "value" and this will be driven by the purpose of the analysis or the questions being asked by the analyst. The valuation of tangible returns (i.e., products and benefits) is different for commercial and non-commercial agricultural operations. This is because the goals, objectives and customers of these operations are very different. These differences require that each has a unique set of metrics to measure value.

As mentioned earlier, the main objective of commercial agricultural operations is the conversion of the crop to currency. *Crop futures* are one way that tangible returns from commercial agricultural operations are valued. Futures contracts are agreements between parties that consent to buy and sell an agricultural crop for a pre-determined price, usually before the crop is harvested; the crop is delivered and payment received at some future date. For crops that are not traded through futures contracts, the contemporary wholesale or retail market price may set the crop's value.

To the commercial operation, assessing the monetary value of the current crop is perhaps the most common method used to assign worth to the harvest. This is a straightforward exercise that can be carried out using available market and sales data to determine what other similar products are currently being sold for. Out of economic necessity or in the absence of sufficient market data, the grower may assign a value to the crop based on production costs and the profit margin that must be realized. The amount of money that the commercial farmer receives through the sale of harvest is referred to as the *cash receipt sales*. It is important to point out that cash receipt sales reflects what the farmer makes from the sale of his harvest. But, other price values exist along the supply chain from the farmer to consumer. If the farmer sells directly to the public, then the farmer can charge a greater price (retail price) for the product than if it were sold to intermediates between the farmer and consumer.

After the farmer sells the crop to the supply chain, *wholesale pricing* is used; wholesale prices are those usually paid by retail stores, restaurants and institutions before the produce is consumed. The wholesale price includes the farmer's sale price plus a sometimes-lengthy list of add-ons associated with transportation, handling by intermediates (e.g., brokers, suppliers), storage, losses during transport, overhead, etc. The consumer's cost (*retail price*) is that which is paid at the grocery store, the retail market or is included in the cost of food served in institutions or restaurants.

Non-commercial operations are usually not preoccupied with the monetary value of what they produce. For those operations, the value of the harvest lies elsewhere- such as how much the grown food reduces hunger in the community or how much the produce reduces the amount of money needed to be spent on food by a household.

For non-commercial agricultural operations, the production and use of the tangible returns are often dissociated from the market system. For example, a home food gardener may grow and consume food without regard for the monetary value of that food. Similarly, a community gardener may grow a bed of flowers to be given away to others as gifts or may be used to decorate headstones in a cemetery. Earlier in this chapter, it was noted that for non-commercial operations (e.g., home food gardens, some community gardens), the financial benefit realized from agricultural production is the amount of money not spent to purchase those crops from a commercial market. For these crops, one may determine a value for them from retail or wholesale sources that sell the product.

Quantifying Intangible Returns

In the previous section, it was noted that people grow food because they receive some kind of valuable returns (products or benefits) from the activity and these returns differ according to the type of operation. As it is with tangible returns, intangible returns can be grouped by those that are directly realized by the farmer (internal to the food growing operation) and those that mostly lie outside of the farming operation (external to the food growing operation). Examples of internal intangible returns include the enjoyment of the gardening activity and a feeling of connectedness with one's food growing heritage. External intangible returns include a community garden's stimulation of community interaction.

Identifying Intangible Returns

The scientific literature, particularly that from the psychological and social sciences, includes a number of studies on the positive effects of growing food. Most of these studies are related to non-commercial agriculture, although those conducting commercial farming operations also realize some types of intangible benefits. Some of the many intangible benefits that have been reported by urban food growers are shown in Table 13.

Psychological

Researchers have documented numerous psychological benefits from interaction with food gardens. Food growing also benefits the community in unexpected ways. Although these intangible benefits may not be easily quantified, nonetheless they are important to measure to understand the magnitude and importance of these benefits. The psychological benefits of urban agriculture can be loosely grouped into classes, which are described below.

Personal Enrichment and Well-Being

Urban gardens and greenspace play an important role in the lives of city dwellers. Gardens, parks and bucolic natural landscapes are used by urban residents to restore a sense of self, to separate from other people, remove themselves from daily rituals and to engage in exploratory behavior (such as observations of nature) (Rubinstein 1997, McConnel & Walls 2005). Byers (2009) described the important role of backyard gardens to immigrants, who reported that gardening provided better health through exercise and stress reduction. Gardens have intrinsic value for the gardener by providing a sense of calm, a place of retreat and a space in which to regain equilibrium. Some researchers have described gardens as places where people can go for solace and emotional relief (Francis & Marcus 1991) and to regain a sense of happiness and mental health (Bliatout 1986). Wilhelm (1975) studied gardening

Table 13. Intangible benefits associated with urban agriculture.

	Commercial Agriculture		Non-commercial Agriculture	
	Internal	External	Internal	External
The field/garden as a destination	x		x	
Beautification		x		x
Enjoyment, relaxation, satisfaction, etc.	x		x	
Connection with one's heritage	x		x	
Exercise opportunities	x		x	
Access to fresher food	x		x	
Healthier food choices			x	x
Greater food variety			x	x
Grow food types associated with cultural preferences			x	x
Encourages community interactions			x	x
Encourages awareness of and protection from local pollution			x	x
Connection with the land	x		x	
Knowledge about where food comes from			x	x
Knowledge about how food was grown			x	x
Knowledge about food safety			x	x
Increases community sustainability (if locally grown & sold)		x		x

within a southern black community and found that residents identified curiosity, personal satisfaction, social recognition, beauty and amazement by the mysteries of nature as some reasons for gardening.

Gardening can provide a sense of fulfillment and accomplishment (Debertin 2012). A study of urban residents found that attitudes toward home food gardening were overwhelmingly positive (Zahina-Ramos 2013). The positive feelings about home food growing were the same for those who belonged to gardening organizations and those who did not, suggesting that these feelings were not related to the group activity itself.

Negative experiences (such as disappointment and frustration with the gardening experience) were associated with gardening failures, such as low crop yield, plant pests and diseases, and frustration because of a lack of gardening skill or knowledge (Zahina-Ramos 2013). Some gardeners expressed a more utilitarian view of their gardening efforts. Although intangible benefits were important, gardening was done primarily to raise the homeowner's standard of living and to provide food (Wilhelm 1975).

In a study conducted in Dunedin, New Zealand (Freeman *et al.* 2012), participants were interviewed concerning residents' feelings about their gardens and their level of knowledge about the plants they grew. Results from this study found that respondents' gardens were extremely important to them and they felt that the garden provided health and healing properties. Among the most frequently reported benefits of the garden were stress reduction, wildlife watching, a sense of ownership, relationship building and expression of self-identity.

In one study, the perceived benefits of urban food gardening varied by age. Middle-aged and older interviewees valued gardening because it made them feel satisfied (when they produced food for themselves and others to enjoy), relaxed and connected with their heritage, including memories of loved ones (Zahina-Ramos 2013). Young interviewees never identified these as values of food gardening. Relf *et al.* (1992) found that younger people were significantly less likely to feel appreciation for the intangible benefits of gardening, including feelings of satisfaction, peace, tranquility, calmness and relaxation.

Urban food production can also reduce consumers' isolation from their connection with the land (Pothukuchi 2004). Freeman *et al.* (2012), after reviewing the literature on the value of intimate contact with nature, stated that living in greener environments has been associated with improved health and vitality (De Vries *et al.* 2003, Nielson & Hansen 2007, Ryan *et al.* 2010), and that contact with plants can be beneficial (Unruh, Smith & Scammel 2000, Tse 2010).

Cultural Expression and Connection with Heritage

Hall (1996) described seven generally recognized values that perpetuate the practice of backyard gardening (both food and non-food): money savings, food freshness, food quality, personal enrichment, a focus for family activities, aesthetics, and community connections. Hall also suggested two other motives: 1) gardening represents a glance back at America's agricultural roots and is culturally important, and 2) it is chiefly an expression of ritual that contributes to the gardener's charting of self and society on a plot close to home.

Gardening lies at the intersection of nature and culture, personal values and public expectation (Kiesling & Manning 2010) and are a work of, and an expression of, culture that reflect the characteristics of the gardener (Lewis 1993). The food plants within a garden reflect the gardener's cultural traits and culinary preferences. Gardens are products of social, physical and symbolic ordering of private living space (Kimber, 2004) and any place where people garden, be it associated with a house or a community garden, is a place of religious, social and cultural significance (Francis & Hester 1990, Kimber 2004, Sinclair 2005).

A study conducted by Zahina-Ramos (2013) suggested that emotional responses to gardening were often valued more than the food that was produced and psychological benefits extended well beyond the growing of plants. Although food gardening was undertaken by study participants to supplement the household's food supply, food was typically grown to provide better quality and culturally important varieties of produce, and to maintain connections with their heritage. These finding are consistent with other studies that found intangible benefits (e.g., enjoyment, relaxation, recreation, detaching from stress, connection to the land and connection with one's heritage) were significantly more important to urban gardeners than other tangible benefits such as food provision (Kaplan 1973, Collum 1995, Dunnett & Qasim 2000, Clayton 2007, Okvat & Zautra 2011, Loram *et al.* 2011, Freeman *et al.* 2012). Vanhonacker *et al.* (2010) and Zakowska-Biemans (2012) described how food cultures were interwoven with regional and national identities, celebrations and uniqueness.

The types of food plants grown within backyard gardens reflect the homeowner's food culture and provide types of herbs, fruits and vegetables that are important to the resident's cuisine. McIlvaine-Newsad *et al.* (2008) and Cobb (2012) noted that food heritage, which is passed through generations, is often reflected in the characteristics of home food gardens, community supported agriculture and hobby farms. Furthermore, gardens create space for social engagement and provide a means for individuals, especially those not native to an area, to maintain parts of their culture by recreating familiar landscapes and cultivating culturally relevant foods. Memories of gardens, particularly those of loved ones', can be powerful intergenerational images (Fivush, Bohanek & Duke 2005). In Byers' study, and in those by Wilhelm (1975) and Møller (2005), the community's elders expressed concern about the younger generation's apathy towards food gardening and feared that the younger generation would abandon the practice of food growing and would lose that valuable connection to their heritage.

Food and Nutrition

Studies have suggested that the presence of farmers markets, community gardens and backyard gardens improve community nutrition and diet (McCormack *et al.* 2010). This could be related to a greater ease of obtaining fresh fruits and vegetables, as suggested by Steptoe *et al.* (1995), Worsley *et al.* (1995) and Nijmeijer (2004). Education-based gardening programs also appear to positively influence participant's receptivity to eating fresh fruits and vegetables. Educational food growing programs' influences on urban youth (Lautenschlager & Smith 2006), grade school children (Linebergert & Zajicek 2000, Skelly & Bradley 2000, Koch *et al.* 2006, McAleese & Rankin 2007, Blair 2009), high school students (Beecha *et al.* 1999) and urban gardeners (Alaimo *et al.* 2008) have been the focus of much research. These studies showed more positive attitudes toward including fresh vegetables in their diet after participating in the food growing programs. Students also had more positive attitudes toward eating fruit and vegetable snacks: female students and younger students had the greatest improvement in snack attitude scores. Although school gardening programs improved students' attitudes toward vegetables, some studies found that improved attitude did not always translate to higher personal fruit and vegetable consumption (Linebergert & Zajicek 2000).

Other benefits of urban agriculture that have been recognized by investigators include:

- Improved dietary choice (Byers 2009)
- Greater food varietal choice and the ability to grow culturally/ethnically-important produce
- Knowledge about where food comes from (Pothukuchi 2004)
- Knowledge about how food was grown (Pothukuchi 2004)
- Knowledge about food safety (Pothukuchi 2004)
- Greater local control over food choice and quality (Pothukuchi 2004)

Community

Community leaders have identified food insecurity and food deserts, places where little fresh or quality food is available for purchase, as growing problems in inner-city neighborhoods. In these areas, access to healthy and affordable food can be limited. Inadequate access to fresh and nutritious food has been linked to higher rates of diabetes, obesity, cardiovascular disease, certain types of cancers and chronic illnesses (McCormack *et al.* 2010, Corrigan 2011, Aubrey *et al.* 2012). Urban agriculture places food production proximal to areas of greatest need, thereby providing a relatively low-cost solution to a social problem (Pothukuchi 2004)

Local food production benefits cities by: 1) providing socio-educational functions, 2) contributing to urban employment, and 3) reducing social inequality (Aubrey *et al.*, 2012). Studies have linked urban food production with economically, environmentally and socially sustainable communities (Mendes *et al* 2008, Sonntag 2008). Other positive impacts extend to areas of nutrition, health, entrepreneurship and social equity (Smit & Nasr 1992).

Freeman *et al.* (2012) found that although gardening has some elements of a solitary activity, some study respondents in New Zealand gave examples of how they use the garden space to create and support new relationships. These respondents used the garden as a point of communication with family members and neighbors.

Private gardens have been found to be important points of social contact and relationship building (Kawane, Tetsuya & Shoichiro 2000) and facilitated interactions among neighbors and family members.

Measuring Intangible Benefits

Intangible benefits are sometimes difficult to measure, especially with the kinds of quantitative metrics that are used in the hard sciences. Social scientists have developed qualitative, semiquantitative and quantitative methods to measure intangible factors that are important to many aspects of human life. Some of these methods include questionnaires, surveys, observational studies, population sampling and controlled experiments (Bernard 2011). Many of these methods allow for statistical analysis and modeling, provided that a sufficiently robust set of data can be obtained. Examples of some methods include (Bernard 2011):

- *Questionnaires*, which can include answers that can be placed along a scale or indexed. Questions can be provided with response options to choose from (categorical, ordinal or interval responses) or the respondent may provide unstructured answers. Questionnaires can address preferences for or attitudes about an activity, practice or item.

- *Informal interviews* are characterized by a total lack of structure or control over the interview- it just flows as a result of natural conversation and there is no clear plan by the interviewer about what types of information are to be collected.

- *Unstructured interviews* are conducted with minimum control over people's responses while keeping a clear plan in mind as to what is to be accomplished; the same respondent may be interviewed on multiple occasions.

- *Semistructured interviews* follow an interview guide that includes a list of topics and questions that must be covered in a particular order; the semistructured interview is best used when the interviewer may only get one chance to speak with a respondent.

- *Structured interviews* follow a strict schedule and list of questions that are administered to the respondent in a way that reduces variability between interview administration.

- *Participant observation* studies allow the investigator to observe and record information about the study participant's life and behavior through either direct observation or indirect observation.

Questionnaires and interviews can be administered through a number of different modes. Interviews are often conducted face-to-face, which can occur while the interviewer and interviewee are in the same space or communicating through on-line chat. Questionnaires can be administered in person (pencil and paper or electronically), via email or through websites. Several websites offer a free or low-cost service that will set up a questionnaire and email it to potential participants. Questionnaires and interviews may be administered to a population at-large to gauge the general attitudes or perspectives about a certain topic. They can also be applied before and after an experience (e.g., participation in an urban food growing effort) to understand how the experience changed participant's feelings about a topic.

Regardless of the method used to explore and document the various intangible benefits realized by urban food growing operations, it is important to measure both the types of benefits and the relative importance of those benefits to the study participants. This way, semi-quantitative and quantitative analyses can be conducted to characterize how urban agriculture benefits city residents.

Indirect Costs and Benefits of Urban Food Growing Operations

Just as there are direct tangible and intangible effects resulting from a food growing operation, there are also indirect or extended effects. These indirect effects are less obvious and more difficult to quantify. An example is the long-term impacts of climate change resulting from the burning of fossil fuel, such as gasoline. Another indirect effect is the methane emissions that arise from burying food waste in landfills. A listing of other indirect costs and benefits that may result from urban agriculture operation are shown in Table 14.

Table 14. Indirect effects associated with urban agriculture operations.

Indirect Benefits
Increase cycling of money within the local economy
Encourages local entrepreneurship
Encourages the development of local and specialized agricultural knowledge
Creates a centralized location for local food growing knowledge
Excess harvest can be more easily diverted to places of need
Reduces non-renewable consumption as compared to food obtained from distant locations
Reduces greenhouse gas emissions
Encourages recycling of waste materials
Provides the community with an alternative disposal option for compostable organic waste material
Outdoor food growing areas provide greenspace for groundwater recharge
Urban gardens can provide habitat for migratory birds and invertebrates
Increases a sense of community self-sufficiency
Provides for a more diverse array of community activities and interactions
Well-maintained sites can contribute to community aesthetics
Responsible water use may decrease urban water demands, especially when the operation replaces a more water consuming land use type
Indirect Costs
Increased risk for groundwater pollution when unsafe practices are used
Importing materials, tools and machinery can increase the urban ecological footprint
Cultivation of invasive species may cause them to escape and proliferate in natural areas
Unsightly or unmaintained sites can decrease community aesthetics
Irresponsible irrigation exacerbates water supply issues and consume valuable potable water

Chapter 9. Environmental Resources and Urban Ecology

In earlier chapters, the effects of environmental factors on agricultural production and the ways that agricultural operations can impact resources were discussed. In this chapter, the ways that food-growing activities can affect environmental resources and urban ecology are explored as well as methods for measuring these impacts or benefits. Agricultural operations may affect environmental resources and urban ecology in direct and indirect ways, and understanding how these operations may affect the local ecology and environment is key to adjusting growing methods to avoid negative impacts and enhance the urban environment. Some environmental impacts from farming operations include eutrophication of water bodies from agricultural runoff, soil erosion and pesticide pollution (Pimentel *et al.* 2005). Urban food production may also produce benefits to the local environment or ecology. These are also important to understand and measure.

Environmental Resources

The cultivation of and adherence to an environmental protection ethic is easiest when people are in contact with the resources that sustain them and when they understand the importance of those resources to their well being. As food production sites have been located further from the point of consumption, consumers have become more isolated from their connection with the land and knowledge about food production, food safety, the resources consumed and the environmental impacts resulting from agriculture. Through time, apathy toward responsible resource consumption, CO_2 emissions, agricultural pollution, pesticide use and food safety issues wanes (Pothukuchi 2004). Urban agriculture promotes pollution awareness within cities (Shutkin 2000, DeLind 2002) and also provides a means to recycle or assimilate waste products from other urban processes.

Urban food growing operations can affect the environment through the consumption of essential inputs (water, energy, fertilizer and other resources) and the creation of waste products. The impacts associated with the use of inputs and waste products vary widely by growing method. Quantification of these impacts is necessary to understand how sustainable a particular operation is and how an operation could become more sustainable in its practices.

Water Resource Use and Conservation

Freshwater supplies are quickly becoming a limited resource in some regions and in other areas water is being consumed faster than it can be replenished. Water usage by agricultural operations is of particular importance for two reasons: to prevent over-use of the resource and reduce food production expenses. In urban areas, irrigation water may be obtained from a number of sources (Table 11) and the amount of water consumed by a food growing operation can be reliably measured. However, what the use of that water means to the environment must also be understood and, if possible, measured to obtain a fuller characterization of the potential impacts or benefits of a food growing operation.

Not all food-growing operations use water in the same quantities and in the same ways. Some operations, particularly those with soils that have a high water holding capacity, apply compost to retard evaporation from the soil surface and use drip irrigation, are efficient water users. Other operations use high-pressure irrigation spray heads and application methods that have lower water use efficiency. In contrast, hydroponic systems use water as the primary growing medium and water must be constantly replenished to replace that which is lost by evaporation from the system.

Overall, urban agricultural operations are more compact than rural commercial and industrialized farms, which can give greater control and flexibility in the type of irrigation used and the amount of water consumed. It can also give more choices with respect to the source of irrigation water. Potable municipal water is generally the most expensive water available to food growers because users must pay a rated fee and associated taxes to use that source. In addition, potable municipal water has undergone treatment and purification processes, which increases the ecological footprint of the water supply. Non-potable municipal water (reuse water) is often used to irrigate landscape and greenspaces and, unless safety can be assured through testing, is not typically used to irrigate food crops. The benefit of using non-potable municipal water supplies is that it is a way to reuse a waste product from another urban process.

Environmental impacts related to municipal water withdrawals and infrastructure vary from place to place, but are significant in large metropolitan areas where natural freshwater supplies are often limited. Some impacts from municipal water use have included unsustainable depletion of aquifers, permanently or seasonal lowering of groundwater and surface water levels, regional ecological impacts and saltwater intrusion into coastal aquifer systems. Although water use regulation laws have been enacted to avoid some of these environmental problems, delays in the observation of impacts and the willingness to accept some level of environmental degradation from municipal water use dilute the level of environmental protection that can be assured by governmental oversight. Other environmental impacts result from the installation and maintenance of a water distribution network, the use of water treatment chemicals, disposal of waste products (e.g., brine from reverse-osmosis or waste products from filtering systems) and the consumption of non-renewable energy to operate water treatment and supply components.

Self-supply wells and surface water withdrawals can offer an inexpensive source of water to growers. In many areas, a water use permit for commercial operations is necessary before the groundwater or surface water resource can be used. However, once obtained, irrigation costs are generally associated with the initial expenditures of piping, pumps and electrical wiring. On-going costs include electricity to run the pumps and systems maintenance. Environmental costs involved with self-supply water include the extended or indirect costs resulting from manufacture of the equipment and supplies used. Direct environmental costs may arise if water is being extracted in quantities that cause or aggravate an unsustainable use of water resources, which can result in impacts to natural systems (e.g., wetlands).

Measuring how a food growing operation's water use may contribute to extended negative impacts (or benefits) is more difficult to quantify. In some regions, there is little data and few, if any, studies that demonstrate a quantitative link between water withdrawals and environmental impacts. In these areas, estimates must be made based on best available information or relationships derived from areas with a similar geologic and hydrologic composition. These estimates should be vetted through local experts and margins of error delineated so that reasonable conclusions can be drawn from these numbers.

The assessment of the environmental impacts to water resources caused by an agricultural operation can be complex. However, it may not be necessary to derive an estimate with a high degree of precision. Instead, a reasonable approximation of impact or benefit that informs the analysis may be sufficient. To begin, one must measure the amount of water being consumed by the food growing operation. Measuring water use is straightforward and many urban agricultural operations have the ability to track water use and sources.

A food growing operation's relative contribution to regional water supply issues can be reasonably estimated once some basic information about the relevant water resources is known. This information can be obtained from a local or regional water supply authorities, who can also provide estimates of the monetary cost involved with avoiding, mitigating or responding to a water supply issue that has been caused by overuse of the resource. This monetary cost can be pro-rated to the agricultural operation as an environmental cost (if the operation is contributing to the overuse problem) or as an environmental benefit (if the operation is helping to alleviate the overuse problem). Pro-rating can be based on the amount of water used by the food grower.

Another way to determine the relative water use cost or benefit from a food growing operation is to compare different land use types. In some regions, freshwater withdrawals to supply urban residents has reached or exceeded the limit of sustainable use. There are concerns that the addition of food producing operations in or near these cities will exacerbate the problem of limited water supplies. Growing food in cities could potentially create additional demand on potable water supplies if these operations are tapping into the same stressed resources. One way to calculate the relative cost or benefit of a food growing operation on water resources is to compare the amount of water typically consumed per year with other land use types. The operation is providing a net benefit if it is consuming less water than an alternate use for the land. For example, it is possible to compare how much water a 5-acre urban farm consumes and compare that volume with the water use demand generated by an urban development of the same area footprint. This approach to the analysis allows one to rank alternative land use types according to their water demand and to see how urban food production may be a more desirable option with respect to water resource use. Quantification of the monetary value of the net benefit (or net cost) can be done using the method outlined in the previous paragraph.

Another example of how urban agriculture may provide a beneficial offset of water use is when the food producing operation replaces a less water-efficient land use, such as irrigated lawn. Urban agricultural land produces many more benefits for the water investment than the irrigated lawn does (many of which are quantifiable and can be monetarily valued). When urban agriculture can provide a net reduction in urban or regional water use (e.g., by replacing a land use type that consumed more water), the food growing operation can provide indirect environmental benefits that lie beyond the city. For example, coastal communities in many low-lying areas are struggling with the problem of saltwater intrusion into freshwater coastal aquifer systems and may benefit by reducing the extent of irrigated lawn. Some of this lawn area may be replaced by productive food growing operations. Local water supply authorities in at-risk areas can provide information about the status of the resource and the costs involved with overuse, including the siting and development of replacement wells if coastal wells are affected by saltwater intrusion. This information can be used to determine the cost or benefits to the water resource realized by the food growing operation, which would be pro-rated as described above.

In urbanized areas, the proliferation of impervious surfaces has disrupted the natural process of rainwater percolation into soils and the recharge of the underlying aquifer. Rainfall runoff is diverted into drainage systems that lead to surface water bodies, reducing the potential for groundwater recharge and contributing to a lowering of the local water table. Urban agricultural lands and greenspace can be important sites for groundwater recharge following a rainfall event. The application of captured rainfall for irrigation further reduces the amount of water withdrawn from the environment to support food-growing operations.

The capture and use of rainwater can be one way to avoid exacerbating overuse of local water resources and can be the least expensive source of irrigation water. If a food growing operation has access to a sufficient area of impervious surface, rainwater can be collected and used as needed. In areas that produce adequate rainfall during the growing season, rainfall collection and distribution systems can be gravity fed and require little operation cost after installation. Environmental costs associated with rainfall collection systems can be minimal if recycled and repurposed materials are used. For small operations, food-grade drums discarded by the food service industry can be retrofitted for water storage. However, large-capacity and manufacture systems carry a greater ecological footprint because of manufacturing, transportation and installation costs. The use of captured rainfall runoff for irrigation would not only avoid an increase in the water supply demand, it may also reduce water use regionally if the urban grown food reduces the amount of food that is grown using less efficient means of irrigation elsewhere in the region. In this case, the urban agriculture would not increase regional water supply demand because it provides an offset of usage from another alternative.

Energy Resource Use and Conservation

Urban and local agriculture can significantly reduce energy consumption related to food growing and transportation. It has been estimated that the average distance that that food in grocery stores has traveled is approximately 2,400 – 4,025 km (1,500 - 2,500 miles) (Florida Fish and Game Commission & The American Farmland Trust 1995, Worldwatch Institute 2016). The long distance transportation of food consumes a large amount of fossil fuels and is a significant contribution to greenhouse gas emissions and air pollution. The global food system also gives rise to other major environmental impacts, including biodiversity loss and water extraction and pollution (Garnett 2013).

The amount of energy saved by producing food locally, in comparison to importing food from outside of the community, can be calculated by determining where food is being imported into the community from and summing up the total distance the produce (expressed as a volume or weight) would have traveled had it not been grown locally. Many local grocery stores can provide information about where the food they are selling has been grown. For those food types that are normally transported by refrigerated over-the-road semis, the amount of petroleum (diesel fuel) that was consumed to transport this food can be assessed from on-line resources (e.g. U.S. Energy Department) using volume, weight and distance traveled estimates. Other estimates from overseas shipping can be similarly calculated. Examples of these calculations are provided in Chapters 12 and 14.

Urban agricultural operations are generally smaller than rural commercial and industrial farms and rely less on machinery and equipment that run on fossil fuels. Petroleum-fueled machinery and equipment consume non-renewable energy resources, which are direct costs to the operation. Other ways that energy can be directly consumed throughout the food production process include the use of electricity from non-renewable sources to power lighting, irrigation pumps, air conditioning/heating of buildings, etc. Indirect costs associated with using non-renewable energy resources extend to the processes of extraction, purification and transportation to the place of sale. There are other indirect costs associated with the use of non-renewable energy. The extraction of fossil fuels in undisturbed areas causes habitat loss and degradation as clearing, exploration and extraction occurs. The burning of fossil fuels causes the emission of greenhouse gasses and other pollutants that cause degradation of air quality and, in some cases, water and soil pollution. These pollutants can have negative effects on human health. Other indirect costs associated with fossil fuel use extend to the environmental damage caused by oil spills, leakage and the disposal of by-products from the refining process.

For agricultural operations that rely on chemical fertilizers, there is a significant amount of energy consumption associated with the use and application of manufactured nitrogen fertilizers. The Haber process, which is used to create nitrogen fertilizer, uses much energy. About 72% of the energy use on conventional (non-organic) farms is due to the energy embodied in fertilizers and pesticides (USEPA 2008). Chemical fertilizers applied to soil partially denitrify and release nitrous oxide and methane, which are potent greenhouse gases.

Calculation of the energy directly consumed by the food growing operation can reasonably be estimated from fuel (liquid or gas) and energy (e.g. electricity) use measurements or purchase receipts. But, the impacts of burning fossil fuels are multifaceted and more complicated. To begin, there are greenhouse gasses that are emitted by the combustion of the fuel. The U.S. Energy Information Administration (2014) states that burning 3.8 liters (1 gallon) of gasoline (that does not contain ethanol) produces approximately 8.9 kg (19.6 lbs.) of CO_2. The emissions are slightly higher for diesel fuel, which produces about 10.2 kg (22.4 lbs.) of CO_2 per 3.8 liters burned.

The National Research Council (2010) published a study that focused on defining and evaluating the health, environmental, security and infrastructural costs and benefits associated with the production and consumption of energy. These are costs and benefits that are not or may not be fully incorporated into the market price of energy, into the federal tax or fee, or into other applicable revenue measures related to production and consumption of energy. This study was conducted under the direction of the U.S. Congress through the U.S. Department of the Treasury. Below are some of the findings from that study.

Results from the National Research Council study showed that for electricity generated by coal, the averaged aggregated nonclimate-change-related damages were estimated to be $0.032 cents per kilowatt hour (kWh, based on 2007 UDS) (Table 15). This value combined monetized effects on human health, impairment of outdoor scenic vistas, agriculture, forestry and damages to building materials associated with emissions of airborne particulate matter, sulfur dioxide and oxides of nitrogen from 406 coal-fired power plants in the United States. These damages are not equally distributed across the landscape, but are concentrated near areas proximal to the power plants and vary according to emissions variation. For 2030, the damages per kWh at coal plants drop to $0.017 per kWh because of expected changes bought about by regulation and policy changes. This value excludes the effects of greenhouse gas emissions. The average emissions from coal-fired electricity-generating facilities in the United States were estimated to be approximately 1 ton of CO_2 per MWh of power generated.

For natural gas electricity generation, the National Research Council study found that the aggregated damages associated with non-greenhouse gas pollution was estimated to be $0.016 per kWh (2007 USD). This value falls to $0.011 per kWh in 2030. Greenhouse gas emissions for natural gas generation were approximately half that of coal- about 0.5 tons per MWh. The caveats provided for coal-fired power plants (above) also apply to natural gas energy generation.

Values for nuclear power, wind, solar and biomass/biofuels are not available from the National Research Council study. However, concerns about emissions from biomass/biofuel facilities may be similar to those from fossil fuel combustion. Other externalities that were not included include damages from electricity transmission and distribution; these can include potential risk from exposure to extremely low-frequency electromagnetic radiation, visual impacts from the infrastructure itself and loss of property value.

The National Research Council study examined fossil fuel consumption for transportation, which was broken down into light-duty (e.g., passenger cars) and heavy-duty on-road vehicles; combined these account for more than 75% of U.S. transportation energy use each year. That study estimated that in 2005, CO_2 emissions from conventional gasoline burned in a standard passenger vehicle was approximately 552 grams per vehicle mile traveled, the mean cost of harm caused by greenhouse gas emissions was approximately $0.014 per vehicle mile traveled and the mean health and non-greenhouse gas damages from light-duty vehicles (for most

Table 15. Monetized damages per unit of energy-related activity[a] (source: National Research Council 2010).

Energy-Related Activity (Fuel Type)	Nonclimate Damage	Climate Damages (per ton CO_2-eq)[c]			
		CO_2-eq Intensity	At $10	At $30	At $100
Electricity generation (coal) per kWh	$0.032	2 lb.	$0.01	$0.03	$0.10
Electricity generation (natural gas) per kWh	$0.016	1 lb.	$0.005	$0.015	$0.05
Transportation[b] per VMT	$0.012 to >$0.017	0.3 to >1.3 lb.	$0.0015 to >$0.0065	$0.045 to >$0.02	$0.015 to >$0.06
Heat production (natural gas)	$0.11/ MCF	140 lb./ MCF	$0.70/ MCF	$2.10/ MCF	$7.00/ MCF

[a]Based on emission estimates for 2005. Damages are expressed in 2007 U.S. dollars. Damages that have not been quantified and monetized are not included.
[b]Transportation fuels include E85 herbaceous, E85 corn stover, hydrogen gaseous, E85 corn, diesel with biodiesel, grid-independent HEV, griddependent
HEV, electric vehicle, CNG, conventional gasoline and RFG, E10, low-sulfur diesel, tar sands. (See Table 7-1 for relative categories of
nonclimate damages and Table 7-2 for relative categories of GHG emissions.)
[c]Often called the "social cost of carbon."
ABBREVIATIONS: CO_2-eq, carbon dioxide equivalent; VMT, vehicle miles traveled; MCF, thousand cubic feet; E85, ethanol 85% blend; HEV, hybrid electric vehicle; CNG, compressed natural gas; RFG, reformulated gasoline.

fuels) ranged from $0.23 to $0.38 per gallon (average = $0.29 per gallon, 2007 USD). For heavy-duty trucks, such as those used to transport food across long distances, these values were much higher. Mean CO_2 emissions from a diesel-burning Class 8 heavy-duty truck was approximately 2000 grams per vehicle mile traveled, the mean cost of harm caused by greenhouse gas emissions was approximately $0.015 per vehicle mile traveled and the non climate change (including health) damages were approximately $0.62 per gallon (2007 USD). These estimates exclude the following:

- Damages to ecosystem and agricultural crops resulting from hazardous air pollution
- Impact of biofuel production on water use, water contamination and other indirect land-use effects

- Electric vehicle components manufacturing and disposal, which can create conditions of exposure to toxic substances and accidental spills

Heat generation for indoor, year-round urban food production facilities requires energy consumption. Most indoor heating comes from natural gas heaters. The National Research Council study found that median estimated non-greenhouse gas damage caused by natural gas combustion (for heating systems in commercial or industrial uses) was found to be approximately $0.11 per thousand ft^3 (MCF) or about 1% of the price of natural gas (2007 USD). Damages in 2030 are expected to be approximately the same as was estimated for 2005.

Much of the energy discussion above focused on what is consumed by food growing operations. However, urban agriculture often replaces other potential uses for the land (such as residential lawn or public park space) and this alternative use is often landscaping dominated by lawn. The choice to install an urban food growing operation, instead of lawn that is maintained by irrigation, the application of chemical fertilizers and fossil-fuel powered equipment, creates a situation where one option may be a more efficient use of resources than another- a situation that should be considered with respect to costs and benefits realized by the operator or community.

Studies of turfgrass management and greenhouse gas emissions found that turfgrass, over the long term, was a net emitter of greenhouse gasses (Zhang 2013, Gu 2015). This was for several reasons, some being the emission of nitrous oxides following application of nitrogen fertilizers, the emissions arising from manufacture of chemical fertilizers and pesticides, and CO_2 from fuels consumed by lawn care machinery. Although these are not always easy to quantify, one source claimed that care of a 1,349 m^2 (0.33 acre) lawn will, on average, consume 68 liters (18 gallons) of gasoline per year because of mowing, trimming, fertilizing, irrigation and other activities (University of Vermont Extension 2016). Along with this fossil fuel consumption are the environmental impacts of extraction, refinement and greenhouse gas release from combustion.

Waste Product Generation and Assimilation

The global food system makes a significant contribution to greenhouse gas emissions ($\approx$12.5%) with all stages in the supply chain, from agricultural production through processing, distribution, retailing, home food preparation and waste, playing a part (United Nations 2012, Garnett 2013). Other activities include the use of mechanical equipment that runs on fossil fuels, the packaging and transport of goods to and from the farm, and the manufacture of equipment and materials used by the food growing operation. Besides CO_2 and methane, nitric oxide (a greenhouse gas) is emitted from soils following the application of synthetic nitrogen fertilizers. The indirect costs of rising greenhouse gas emissions can be manifested many ways (e.g., loss of agricultural lands, habitat loss, increasing fire frequency in some regions), but the most financially noteworthy may be the large amounts of money being spent by governments to curb emissions- costs with are passed on to citizens through taxes.

Waste products from food growing operations may include plastics (weed block, sheeting for solarizing, hoop houses, potting material, etc.), discarded organic waste and metals (discarded tools and equipment). Waste pickup and disposal costs for a food growing operations in a particular municipality can be calculated from budget data obtained from local waste management providers and pro-rated for total volume collected. In that absence of such data, one must rely upon averaged cost data obtained from other sources or studies. For example, the North Carolina Division of Pollution Prevention and Environmental Assistance (Hunt, Howes & D.P.P.E.A Director 1997) compiled data on the cost of waste disposal from 11 municipalities in North Carolina and found a relationship between the size of the city's population and the cost to dispose of 1 ton of waste (Figure 7). Similar data may be available from other regions and are useful to infer general costs for a specific area.

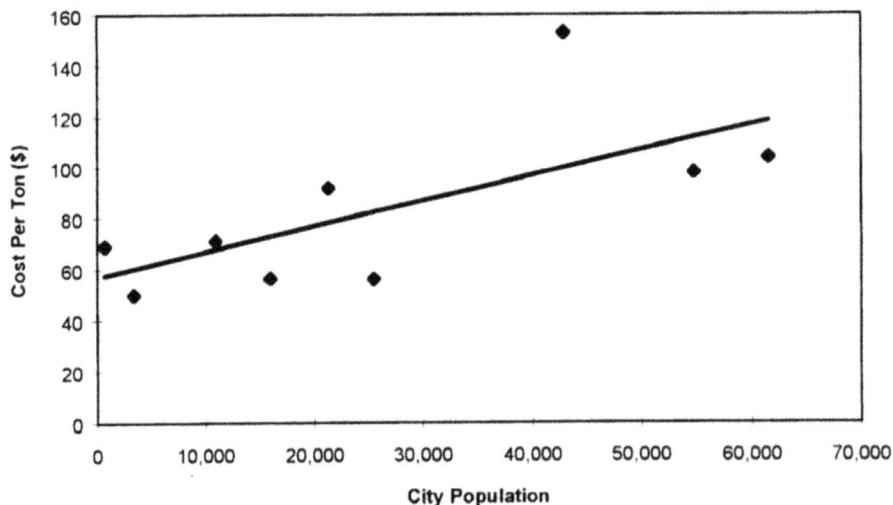

Figure 7. Solid waste costs per population center size in North Carolina, 1994-1996 (figure source: Hunt, Howes and D.P.P.E.A. Director 1997, *n*=11).

The landfills are a significant source of greenhouse gas emissions, particularly methane. Estimates of long-term methane emissions from various materials that are commonly disposed in landfills are useful to calculate the amount of greenhouse gas impacts that could arise from disposal of wastes from a food growing operation. These emissions estimates are available from the US Environmental Protection

Agency's WARM model (US Environmental Protection Agency 2014). Food scraps produce the highest emissions of methane in landfills and recovery of these materials can yield the greatest benefit.

The waste assimilation capacity of urban agricultural lands is significant. Some waste materials (lumber, containers, etc.) can be reused by urban food growing operations as a means to reduce overhead costs to the grower and a convenient place to recycle unwanted materials. This is particularly true for compostable organic waste products that can be turned into a valuable soil amendment. This is a means to recycle nutrients, reduce waste sent for landfill disposal and to reduce greenhouse gas emissions. The net benefit of a food growing operation, with respect to reducing the waste stream to landfills, may be significant.

Plant Nutrient Use, Conservation and Management

Farm operations that rely on chemical (mineral) fertilizers run primarily on inputs that carry environmental costs, some of which carry indirect costs that are often not realized by the grower. Chemical fertilizers consist mainly of inorganic substances that have been mined (e.g., phosphate) or have been created by energy expensive processes (e.g., ammonium). By using a chemical fertilizer on the farm field, the farmer is increasing crop yield but is also altering the soil's chemical composition and contributing to the loss of soil organic matter, both of which can have long-term detrimental effects. When relatively higher amounts of fertilizer are applied to the farm field (more than is taken up by soil microbes and crop plants), excess is lost from the soil. Where this excessive fertilizer ends up can create undesirable consequences elsewhere and these consequences can come with considerable costs to others and the environment.

If the soil at the food-growing site has a high leaching potential (such as sand), some portion of the fertilizer can percolate through the soil column and pollute the groundwater (surficial aquifer) below. Groundwater pollution can have both environmental and human impacts. For example, nitrate (NO-3) is added to the soil by fertilizers or can be produced naturally by soil microorganisms through the mineralization of organic matter (Nemecek *et al.* 2007). Because nitrate is easily dissolved in water, it can percolate into the ground water under certain conditions. Soil nitrate losses are undesirable because this loss of nutrients from the soil increases the need for fertilizer supplements and nitrate contamination of potable water sources may have a toxic impact to humans. Although the acute toxicity of nitrate is low, nitrate is easily converted into nitrite, which has a higher acute toxicity and is supposed to be indirectly carcinogenic (Surbeck & Leu 1998). As pollution levels in groundwater increases, so do the costs of treatment and monitoring.

Nutrient-enriched runoff from fields often ends up in streams or lakes, which can alter the water chemistry and ecological characteristics of these water bodies. Eutrophication of surface water bodies causes ecological changes, contributes to algae blooms, impacts fisheries and degrades water quality. The United States Environmental Protection Agency estimated that half of the pollution in rivers and streams in the United States originated from farm runoff (Windham, 2007; Hoffpauir, 2009). The "dead zone" in the Gulf of Mexico has been caused, in part, by nutrient runoff from agricultural fields throughout the Mississippi River watershed.

Phosphate in agricultural runoff is responsible for eutrophication and degradation of water quality in water bodies and wetlands. However, the lesser-known impact comes from using phosphate fertilizers from inorganic sources. Environmental costs associated with phosphate mining operations include the destruction of natural habitat, a loss of ecological goods and services that the natural habitat provided, generation of waste products and pollution. For example, the island nation of Nauru was once held significant phosphate rock reserves, but mining these deposits for fertilizers and other uses has exhausted nearly all of the island's resources. Today, approximately 80 percent of the land surface of Nauru has been strip mined, which has had catastrophic effects on the island's economy, environment and offshore fisheries with up to 40 percent of marine life killed by downstream effects (Colt 2011). In central Florida (U.S), phosphate mining has destroyed 199,000 hectares (492,000 acres) of natural habitat (Florida Department of Environmental Protection 2017). The per-kg environmental impact of mined phosphate is difficult to quantify, however it should be included in the list of harmful effects that are realized by its use.

Nitrogen (in the form of NH_4^+ or NO_3) fertilizers applied to agricultural crops may originate from two sources: the form of N-fertilizer derived from laboratory processes or from natural processes that transform atmospheric N_2 into plant-usable forms via biological N_2 fixation (BNF) (Peoples, Herridge and Ladha 1995; Nemecek *et al.* 2007). Although BNF has traditionally been a component of many farming systems, its importance as a primary source of N for food production has diminished in recent decades as increasing amounts of fertilizer-N are used for the production of food and cash crops (Peoples, Herridge and Ladha 1995). Chemical nitrogen fertilizers are created through an energy-expensive process that consumes fossil fuels and releases CO_2 to the atmosphere. When these concentrated fertilizers are applied to farm fields, they cause emissions of nitrous oxide, a major greenhouse gas (Bouwman 1996).

Some of the direct and indirect costs of fertilizer runoff into surface and ground water supplies in the United States have been estimated (Union of Concerned Scientists USA 2016); these are:

- The cost of treating private drinking water supplies contaminated with fertilizer pollution was estimated to be approximately 1.12 billion between 1992 and 2004.
- Removing nitrates (from agricultural sources) from municipal water supplies cost approximately $1.7 billion per year, according to the USDA
- Impacts to fisheries from agricultural fertilizer pollution was estimated to be over $100 million per year and related impacts to tourism was estimated to be nearly $1 billion per year
- Respiratory disease associated with nitrogen fertilizers was estimated to cost approximately $23.10 for each kg of fertilizer
- The total cost of fertilizer pollution in the United States was estimated to be approximately $157 billion per year, more than double the value of the 2011 corn harvest

Because of a myriad of variables, it is not easy to estimate the contribution of a single food growing operation (that uses chemical fertilizers) to these reported costs. However, one method is to divide the total dollar value of impacts (the total costs of water treatment, impacts to fisheries and impacts to tourism = $94 billion/year) by total the number of acres of non-organic farmland in the United States. According to the Natural Resources Conservation Service (2016), in 2007 there was approximately 357.0 million acres of farmland in the United States. Of this, approximately 2.285 million acres was organic cropland (USDA Economic Research Service 2016), leaving a total of 354.7 million acres of non-organic farms. The average environmental cost per acre distributed across all of these farms was approximately $266 (2007 USD).

Organic farming operations do not rely on chemical or mineral fertilizers. Instead, they use cover crops to replenish soil nitrogen supplies, as well as compost and manure amendments. Many of these amendments have high organic matter content, which absorb water and increases soil water holding capacity- reducing runoff and infiltration into groundwater. Organic fertilizers (e.g., manure or compost) are usually recovered from other processes, such as waste recycling (Nemecek *et al.* 2007). Because organic fertilizers recycle nutrients from waste products, they can carry less environmental costs than chemical fertilizers. The demand for compost and mulch in food growing operations can play an important role in creating a closed loop system where a waste product from one process is used as a valuable resource in another process. Although this is a real benefit from organic agricultural methods, the quantification of this benefit is difficult to value.

Other Resources

The manufacture and maintenance of agricultural machinery, buildings, tools and other infrastructural components consumes resources and energy is required for resource extraction, manufacturing and transport of both raw materials and finished product. Some of the costs associated with manufacturing, maintenance and operation of agricultural machinery include:

- Materials resources: petroleum products, glass, plastics, rubber, metals, lubricants, finishes, etc.
- Energy resources: fuel oil, natural gas, electricity, coal, etc.
- Maintenance and repair: labor, lubricants, filters, batteries, parts, tires, etc.
- Waste products: emissions to the atmosphere, petroleum byproducts, glass, paper, plastic, rubber, metals, etc.

The costs associated with the purchase of these materials can be tracked from receipts and purchase records. However, the indirect and extended costs associated with the manufacture and transport of these materials is difficult to value.

Ecological Resources

Urbanized lands are a mosaic of hardscape and greenspace, each varying in density along a gradient from the urban core to the outer suburbs. This physical layout shapes many characteristics of the urban ecology. Overall, urban ecology is poorly studied. However, as the world's population continues to grow and urban areas expand, the urban ecological functions need to be studied seriously. The potential role that food production sites can play in raising to urban ecological function must be better understood and appreciated as these can play a significant role in restoring what was lost when natural habitat was cleared for development.

Urbanization and land alteration to accommodate large-scale rural agriculture has profoundly reduced the extent of natural habitat around the world and is a significant factor in biodiversity loss and species extinctions throughout the 20th century. This trend is continuing into the 21st century. According to the United Nations, the worldwide extent of urbanized land is predicted to triple between 2000 and 2040; the land affected by this urbanization will be roughly equivalent to the area of the continent of Africa (Elmqvist 2012). With so much natural habitat already lost to human activities and more anticipated over the next half-century, it is imperative that developed areas and new development adopt measures to support urban ecological functions. Urban agriculture can contribute to this objective.

Food growing lands within cities are part of the urban greenspace and can be places of significant ecological biodiversity, depending on how they are managed and what is grown. Because urban and near urban food producing operations tend to grow a variety of crops, a greater number of insects, birds and other organisms can be found there as compared to an urban lawn or densely developed area (Ryall & Hatherell 2003, Sperline & Lortie 2010). The tendency to increase biodiversity is what sets urban agriculture aside from industrialized rural agriculture, which tends to remove all traces of the natural ecosystem from the growing site. Urban agriculture often begins with vacant land that had been cleared for other purposes and the food crop plantings increase biodiversity on the site it occupies. It is this characteristic of urban agriculture that makes it a potentially important part of the urban ecology and careful management of the growing space can provide significant contributions to the urban environment.

Many agricultural crops rely on pollination to produce fruiting plant parts. Food plant pollination is often viewed as a rural agricultural issue, but the importance of insect pollinators to urban agriculture has received little consideration. In order to support a growing community of urban food growers, a robust insect population that includes pollinators, agricultural pest predators and agricultural pest parasites is essential. Natural and recreated habitats, as well as urban gardens, within developed areas are typically viewed as having aesthetic, hydrologic (e.g., groundwater recharge) and wildlife value. However, they may also support a number of organisms that are important to urban agricultural production. Vacant lots, waste areas and wildflower gardens may also provide habitat for the production and dispersion of beneficial food plant pollinators in cities. The use of toxic pesticides on these areas (and in gardens) can have detrimental effects on not just the urban ecology, but on urban agriculture too.

Urban agriculture can also have negative impacts on the local ecology. Some non-native food plants can escape cultivation and disrupt natural ecosystems. Many food growing operations do not screen plants for invasiveness and agricultural information sources are usually reactive, rather than proactive, with respect to issuing guidelines for growing potentially invasive species. A number of naturalized plant species are escapes from (ornamental and food) gardens.

Some naturalized plant species are the worst agricultural weeds. Weeds, whether in farm fields, roadside swales or vacant lots in cities, can provide important ecological functions. Most weeds are larval food plants for butterflies and moths, produce flowers that attract pollinators and extract nutrients from the soil and concentrate them in plant tissue. Many weeds are not just edible, but delicious and nutritious to eat. Some of these are cherished food plants in other cultures. One example is the Portulaca (*Portulaca oleracea*), which has an almost worldwide distribution. In the United States, it is considered an exotic weed but is eaten in Asia, Europe, Mexico and the Middle East. Purslane also has medicinal qualities, can benefit agricultural crops that it grows alongside of and can be a valuable cover crop (or living mulch) in hot climates with poor soils.

Quantifying Ecological Aspects of Urban Agriculture

The previous sections in this chapter discussed the various ways that urban agriculture is interconnected with environmental resource conservation, resource management and the urban ecology. The interaction between urban agriculture and the environment can yield important benefits and impacts. Standard methods to measure ecological parameters can be applied to the urban food growing space to determine species composition, richness and diversity, as well as the functions that these species can play in the local ecology. Summarizing these data can provide insight into the ecological functions that the growing space provides to the urban ecology. It is preferable to conduct ecological surveys before and after the implementation of an outdoor food growing operation so the pre-and post-data can be compared. Current ecological data can also be compared with other urban and non-urban sites to understand how closely a food growing space's biological components resemble, or are different from, a more natural community. This analysis may be of interest to those who are interested in using growing methods that encourage biological function or diversity that more closely resembles a natural ecosystem.

Urban food producers often rely on materials and inputs that originate from other processes. For example, mulch is waste product from landscaping operations. Other waste products that are valuable resources for some urban agricultural operations include compostable materials, reclaimed wood and reuse water. The volume of material that has been captured and reused (or recycled) can be recorded. Valuation of these materials can be done in one of two ways: (1) the amount of money saved by the operation by reusing materials they would have otherwise had to purchase, and (2) the amount of money saved because the materials did not have to be disposed of. An extended benefit of reusing or recycling materials is the cost to the community associated with disposal in a landfill and the environmental impacts resulting from that fate. Some landfill operations may be able to provide estimated of the latter cost on a per-ton unit.

Besides the measurement of ecological parameters (as described in the previous paragraph), there are other aspects to a food growing operation that overlap the environmental and social parts of sustainability. These are typically more difficult to measure and include:

- Pollution awareness; urban agriculture can sensitize residents about pollution issues in their community, especially if a type of pollution can contaminate the food growing space
- Awareness of where food comes from, its safety and how it is grown
- Understanding how food production consumes resources and causes environmental impacts

It is recognized that these intangible benefits may not be converted to tangible values, however their importance is recognized and need to be reported alongside other costs and benefits.

Chapter 10. Cost-Benefit Analysis for Urban Agricultural Systems

The benefits from agricultural operations have often been estimated from a simple ledger sheet analysis using a limited set of parameters, such as productivity, sales, income and expenses. Other parameters, such as those considered in a life-cycle assessment (e.g., impacts of greenhouse gas emissions resulting from farming operations), are rarely included. These other parameters, as well as intangible costs and benefits, can be significant considerations when deciding on the "worthwhileness" of a food growing operation. Exclusion of important parameters or placing emphasis on a certain perspective (financial profitability) has often led to incomplete measurements of the costs and benefits of an agricultural operation.

The cost-benefit analysis (CBA), which has also been referred to as a benefit-cost analysis by some authors, is a method for systematically and explicitly estimating the relative contribution of different factors and elements to a business process. According to Fuguitt & Wilcox (1999), the "cost-benefit analysis is relevant for those decisions (government or private) which (a) involve the use of scarce resources and (b) generate "good" and/or "bad" consequences for social welfare." The CBA is used to examine a more robust range of inputs and outputs from a business activity or a policy, and may be successfully applied to urban food growing operations, regardless if they are commercial or non-commercial in nature. The value of conducting a CBA for an urban food growing operation is that many intangible and indirect benefits and costs can be recognized, accounted for and expressed in terms of monetary value, where reasonably possible. This exercise can illuminate the range of social, economic and environmental aspects that are of interest to sustainability scientists.

This chapter describes how a CBA can be applied to urban agriculture and how some of the intangible, implicit and indirect benefits and costs may be valued. The methods described in this chapter come from several sources and are by no means complete, but they do provide an overview of the theory behind the CBA and the examples provided are instructive for those who wish to conduct a CBA for a modest operation. When conducting a CBA for a more complex set of conditions than are provided here, it is best to consult an authoritative source.

The Cost-Benefit Analysis: An Overview

Models are used to represent the functions of a system and to simulate how that system will change under a different set of conditions. The CBA is a type of model that can be used to represent the general functions of a system that has economic aspects. The CBA assesses a given policy or activity that uses *scarce resources* and generates benefits and costs through time (Fuguitt & Wilcox 1999). Fuguitt & Wilcox (1999) describe *scarce resources* as "finite resources with alternative uses." Because a food growing operation (or other activity) utilizes scarce resources (water, fertilizers, monetary capital, etc.), these resources are no longer available for other uses. The CBA allows the analyst to compare the efficiency or desirability of using resources for the food growing operation as compared to other potential uses. Cost-benefit analyses have been used:

- To account for and calculate expenditures and benefit values of an activity for the purpose of informing decision makers and managers (Campbell & Brown 2003)
- To assess the comprehensive and distributional impacts of projects and policies (Nugent 1999)
- In a wide range of applications, from vaccination programs (Nichol 2001) to criminal law (Brown 2004)

The CBA is well suited for use in the sustainability sciences. The concept of sustainability is often presented in one of two contexts. The first context is binary- a practice is either sustainable or not. This context assumes that what constitutes a sustainable condition has been defined and a practice is compared with this definition to determine whether or not it meets the specified criteria. The second context that sustainability is often framed in references a practice's position along a gradient from "not sustainable" to "highly sustainable". With respect to the role of a CBA in the sustainability sciences, the CBA may be best applied as a determination of whether some operation, practice or policy is "more sustainable" than another.

The CBA procedure and valuation methods can be successfully applied to urban food growing operations in the following ways:

- Deciding if a proposed operation is advisable- can enough benefit be realized to make the investment worthwhile?
- Assessing the desirability of alternative operational scenarios or policies without actually implementing them (Debertin 2012)
- Examining the workings of an existing operation to understand its profitability and net benefits
- Understanding the future value of the operation

Nugent (1999) proposed a CBA to identify and quantify the economic, social and ecological impacts of growing and distributing food to urban food consumers. This CBA approach was a way to understand how agricultural production in cities changed urban food systems with respect to the three major categories of sustainability (i.e., economic, social and environmental). Nugent proposed a list of benefits from urban agriculture that should be included in a CBA, including:

- Agricultural production (marketed and non-marketed)
- Indirect economic benefits (multiplier effects, recreational opportunities, economic diversification/stability and reduced disposal of wastes)
- Social and psychological benefits (food security, dietary diversity, personal psychological benefits and community well-being)
- Ecological benefits (hydrologic function, air quality and soil quality)

Costs of urban agriculture were identified as inputs and returns. Inputs included natural resources (land and water), labor (wage and voluntary labor) and capital/raw materials (tools and machinery, fertilizer and pesticides, seeds and plants, and energy). Outputs identified included pollution and wastes, such as waste disposal and soil, air and water quality impacts. Although Nugent stopped short of conducting a CBA for urban food systems, the proposed framework is useful to identify the different inputs and returns required for such an analysis.

One of the difficulties in conducting a CBA is the problem of valuating intangible factors. Although it is ideal to have quantitative data input for expenditure and benefit values, this is usually not the case and a number of techniques have been developed to fix quantitative or semi-quantitative values on unknown or intangible factors (Fuguitt & Wilcox, 1999). The CBA is based on valuation theory, which assumes that all things can be assigned a reasonably accurate value (Adler & Posner 2001). With respect to urban agriculture, some of the intangible costs and benefits can include:

- Contribution to regional food sovereignty
- Reduction of food insecurity
- Reduction in the extent of food deserts
- Increased self-sufficiency
- Enjoyment
- Personal satisfaction
- Frustration
- Commitment to crop production
- Connection with one's heritage

- Knowledge about how food was grown, where it was grown and its safety
- Provision of fresher produce
- Use of the food growing space as a social focal point
- Contribution to economic stability

Some methods that have been used to place values on intangibles include community surveys of value perceptions, risk assessments, willingness to pay for services or benefits, and cost-effectiveness analysis. These can be fraught with controversy when used in inappropriate ways or applied to social or human health contexts (Fuguitt & Wilcox 1999, Adler & Posner 2001). Fuguitt & Wilcox suggested that when the benefits from an activity have hard-to-measure effects, the analyst should: (1) value as many benefits and costs as possible using monetary units; (2) if unable to assign a monetary value, try to quantify it in physical units; or (3) describe the benefit qualitatively if the other options are not possible. The analyst should always explicitly identify values in the analysis that have not been assigned so that stakeholders or managers can integrate that information into their determination of the operation's or policy's value.

The CBA Framework

The CBA Setup

The CBA is conducted by an analyst, or a group who functions as an analyst, and is usually developed in an electronic spreadsheet program. The analyst is responsible for defining the purpose of the CBA, determining what elements will be included in the analysis, outlining how these elements will be valued, determining how unvalued elements will be treated and reporting the results. The analyst is also responsible for identifying all consequences associated with a proposed activity, determining all of an operation's or policy's impacts that have good or bad consequences and for outlining these consequences to the decision maker or operations manager. A CBA should provide a comprehensive assessment of an operation's or policy's effects on the investors, the referent group and society. For these groups, the analyst should identify areas where benefits and costs could occur and represent the economic value of changes associated with the activity. Effects of any proposed operation or activity can be divided into two categories: the first are those that are appropriately represented by market prices (marketed) and those that are not (external).

The CBA is usually conducted with respect to a *referent group*, which is usually defined by the client (i.e. the person who is ordering the analysis and defining parameters) or decision maker. In short, it is the group who is relevant to the analysis. An urban agriculture CBA can be conducted for a number of different referent groups from different levels of scale. One referent group could be a household; the CBA would inform them about how a backyard food garden would benefit them and their community. Another referent group could be a municipality who would like to invest and encourage food growing operations for the purpose of enhancing sustainability initiatives, reducing strains to groundwater supplies and municipal water production, increasing food security, increasing community interaction and retaining more monetary capital within the community. Still another referent group could be a state or federal agency that wishes to evaluate the efficiency of a new program to support and encourage urban agriculture projects at local levels. In this latter case, the amount of energy (fuel) savings related to reduced food transportation (an issue of interest to national energy policy), regional environmental benefits (an issue of interest to species diversity protection and water resources protection) and national food security are issues that the CBA could examine. Each of these referent groups is interested in the CBA of urban agriculture, but approach the analysis from very different perspectives and needs.

The referent group may also be defined as all groups affected by the project (the stakeholders). For an urban farm, the referent group may be the farming operation's business manager. For a residential food garden, the referent group is the head of the household. The types of inputs, outputs, benefits, costs, investments, profits, taxes, interest, debt costs and processes considered in the CBA are based on the referent group's perspective of what is relevant and important. The inclusion of environmental and societal effects may or may not be included in a CBA, depending on the purpose of the analysis, but these are required considerations for any CBA that deal with sustainability issues.

The CBA Structure and Calculations

One of the first steps in developing a CBA is to identify all known positive and negative consequences or results that may be generated from a proposed (or on-going) activity (Fuguitt & Wilcox 1999). These consequences or results can be financial loss, profit, environmental impacts and unintended social benefits. It is often necessary to consult outside sources of information to fully understand what the extended or unintended impacts may be. For example, professional hydrologists who are familiar with the region's hydrogeology can provide valuable insight about how surface or groundwater withdrawals by a food growing operation may contribute to water resource depletion.

The CBA usually calculates net benefits through a pre-determined period of time. For food growing operations, benefits are normally calculated for a yearly time step lasting from a few years to a decade or more. The analyst will fix the CBA time period (number of years the analysis spans) according to limiting factors. For example, if a food growing operation has a land lease that lasts 7 years, the analyst should restrict the CBA to the duration of the lease period. A CBA that lasts longer than a decade may be unreasonable in some cases, particularly where the presence of significant uncertainties may render long-term predictions of operations unreliable.

The basic structure of a generic food growing operation's economic CBA is shown in Figure 8. All inputs, outputs, costs and benefits are expressed in terms of a common currency. Throughout this book, monetary values will be expressed in U.S. dollars (USD) except where otherwise noted.

PROJECT NET CASH FLOW					-----YEAR OF OPERATION-----							
		Initial	FY-01	FY-02	FY-03	FY-04	FY-05	FY-06	FY-07	FY-08	FY-09	FY-10
Investment Costs												
	Fixed investment	-$15,000										
	Working capital	-$3,000										
Total investment		-$18,000										
Operating costs			-$12,000	-$10,000	-$7,000	-$7,000	-$7,000	-$7,000	-$7,000	-$7,000	-$7,000	-$7,000
Revenues			$12,500	$15,000	$17,500	$20,000	$20,000	$20,000	$20,000	$20,000	$20,000	$20,000
Net Cash Flow		-$18,000	$500	$5,000	$10,500	$13,000	$13,000	$13,000	$13,000	$13,000	$13,000	$13,000
		1%	3%	5%								
Net Present Value (NPV)		$81,665	$68,862	$58,154								
Internal rate of return of investment (IRR)		39%										

Figure 8. Sample economic cost-benefit analysis structure.

The CBA places all (historic or future) earnings or losses into the same time frame using a *discount rate* by applying a percent change in the value of currency over time. This is necessary because the value of a dollar earned a year from now will most likely be different than a dollar earned today. By applying a discount rate to each time step in the analysis, the calculated dollar values are placed into a common reference point with respect to value and are, therefore, comparable. Another adjustment of monetary values within the CBA is the calculation of an *opportunity cost*, which considers potential lost opportunities to invest the money (used to start and run the operation) in another less-risky venture. The opportunity cost can be explained in the following example- a dollar earned today could have been invested and interest could have accrued through a low-risk investment alternative such as a savings account at a commercial bank. The dollar spent today has less value in the future because the potential future return on investment of that dollar has been lost (Debertin 2012).

Because the actual change in currency value can be uncertain, a bracket of ranges is often applied to calculate different discount rates and to understand how sensitive the analysis is to currency value fluctuations (Figure 8). The reference time frame is usually the present and the net value of costs and profits that are calculated for a given time step is referred to as the *net present value* (NPV). Note that the CBA shown in Figure 8 only contains internal costs and revenues (i.e., those realized by the operation) and excludes external effects such as impacts from pollution runoff.

The generic CBA has several parts:

- Startup costs, which includes fixed investments, working capital and other items described in Table 7
- On-going costs such as operational expenses (Table 6, Table 7 and Table 8), investments, revenues, taxes, interest, debt payments
- Some CBAs will include increases or decreases in risk (e.g., environmental degradation or destruction)
- Profits or value increases (revenues) realized by the operation
- Opportunity costs
- Yearly calculations of net cash flow that have been adjusted for inflation values using a discounting rate; discounting is used to define what a specific amount of future revenue or cost would be worth today (Debertin 2012).
- Calculation of the NPV for the entire analysis period
- Calculation of the internal rate of return (IRR); higher IRR values are more desirable and indicate that the activity is worth undertaking with respect to the inputs and outputs that have been included in the analysis

The CBA outline shown in Figure 8 does not include extended (external) costs and benefits, as well as indirect costs and benefits (either tangible or intangible). Food growing operations that are concerned with the sustainability aspects of their activities should include a more robust suite of inputs and outputs. In the language of the CBA, these are referred to as external effects- i.e., effects that occur outside the market system and are usually affecting (positively or negatively) others members of society (Fuguitt & Wilcox 1999). Some of the external costs of agricultural operations can include:

- Greenhouse gas damages resulting from the consumption of fossil fuels
- Non-greenhouse gas damages resulting from the consumption of fossil fuels
- Waste disposal costs
- Damages resulting from using irrigation water obtained unsustainable sources
- Downstream nutrient pollution impacts

The costs identified in the above bullets may be valued according to the methods identified in previous chapters. However, there are other external costs and benefits that may not be able to be valued; some of these are listed in Table 13. These other external effects may be segregated into two categories of value: use value and non-use value (the latter originated from environmental resources CBA). For example, a non-use or existence value is the value of something to people because it exists. Because it is relative and highly dependent on context, it can be difficult to assign a value to or it should be excluded altogether from the analysis. There are other controversies and problems with the traditional CBA; some examples are provided below.

Valuing Hard-To-Measure Effects

Many of the items listed in Table 13 are difficult, if not impossible, to measure. This raises the question of how to handle them in the CBA. Recall that the CBA is both an analytic and a decision making tool. Although it is ideal that all inputs and outputs (effects) in the CBA are expressed as a currency value, this is not feasible. These hard-to-measure effects are dealt with in one of three ways (Fuguitt & Wilcox 1999):

- Benefits and costs are quantified in monetary units according to *willingness to pay*, i.e. how much one is willing to pay for the effect. This may also be extended to how much it costs to undo impacts from an operation. An example of "willingness to pay" valuation can be applied to labor costs- one can use surveys of the labor force to determine what a reasonable estimate of labor value would be for volunteer or unpaid work. However, when "willingness to pay" is based on the cost of reparations, caution must be used. For example, the average expense associated with cleaning up agricultural pollution, as pro-rated by hectare, can be ascertained from industry or government data. However, this underestimates the value of cost since there are other externals that are not included, such as the loss of ecological goods and services during the time when the resource was impaired.

- Some effects can be measured in physical units, such as kilograms of CO_2 emissions. This avoids placing a monetary value on the greenhouse gas emissions while maintaining some measure of effect.

- Where some effects cannot be measured quantitatively, these must be included in the analysis but described qualitatively. Some examples are importance of the activity of growing food to the grower, the enjoyment of growing one's own food and the aesthetic value of the garden.

Issues with the CBA

Exclusion Bias

Exclusion bias arises when one only considers a subset of the significant inputs, outputs, factors and impacts from an agricultural practice, based on what is considered to be most important or what data is available. Exclusion bias, in a sense, is similar to sampling bias in statistics, where some members of a statistical population have a lower chance of being included in a sample or are excluded altogether. Exclusion bias can occur when the analyst uses a narrow perspective about what is considered to be important. For example, a large-scale agricultural operation that is focused on financial success will often define the costs (labor, materials, land taxes, depreciation, etc.) and benefits (profit from products sold) in economic terms. In this case, what is considered to be a successful operation is one that makes a net profit. However, this type of analysis ignores extended costs (and benefits) associated with the operation because they are viewed as being unimportant to the primary goal- profitability. In contrast, small-scale operations, such as community gardens, have a different set of values, perspectives and goals. Generating a net profit is usually beyond the scope of their charter and an analyst examining those operations may value intangible benefits more than financial benefits.

With respect to the above examples, it is clear that results from any of these analyses would not be highly applicable to food growing operations elsewhere and are of questionable academic value. Ultimately, any analysis that has excluded factors that could have significantly affected the outcome cannot be considered worthwhile for any broader application. By doing studies and analyses that more fully characterize urban agricultural inputs/outputs and that quantify a robust set of costs, the risk for exclusion bias is reduced and more accurate characterizations of benefits and harm can be obtained.

Risk Equity

When using CBA as a means of risk assessment, there are some aspects of the nature of risk policy that is usually poorly represented. Risk management policies are typically focused on eliminating risk, reducing risk exposure, mitigating risk or creating a more equitable distribution of risk. This suite of options may not be well represented in the CBA analysis. Typically, CBA analysis often focuses on incremental risk associated with a single risk exposure, rather than multiple exposures at once. Because of this, synergistic effects are not accounted for when several risks may be interacting (e.g. risk of cancer from exposure to a single vs. multiple carcinogens). Another problem with risk assessment is whether it should be concerned with risks people actually face or those they perceive that they face.

Agencies conducting CBA routinely ignore objectively bad preferences (people who dislike something and are willing to pay to deliberately harm it or keep it away from them) and refuse to weight certain kinds of objectively bad preferences. This selective exclusion can bias an analysis.

Valuation Theory

CBA is based on valuation theory or the theory that all things can be assigned some reasonably accurate value. This becomes a problem when dollar values are assigned based on scant or baseless information. Other issues with contrived values in a CBA include the valuation of human life/welfare and comparisons of dissimilar categories, such as the benefits of a power plant versus the power plant's periodic impact on vistas at the Grand Canyon. The CBA attempts to put all relevant costs and benefits into a common temporal and valuation context, but often these values are unmeasurable or based on weak surrogate valuations. For example, "willingness to pay" is often obtained by surveys and used to value some non-market items. But this value can be biased by income level, demographics, how the question was framed and cultural characteristics of the person being surveyed. Sometimes, unrealistic values are obtained from surveys. For example, the willingness to pay for protection from a pollutant may be far greater than the cost to clean it up. Other problems with the "willingness to pay" approach is that the value of a dollar varies between individuals, local economies, scale, cultures, etc. In contexts that involve increasing risks, preferences and perceived values are less rational and more influenced by emotion. Some things are probably impossible to value, such as the choosing of one option leads to someone having offspring that they would not have under another option. How are the benefits or cost of this new future individual on a future society to be valuated?

Nowhere is the use of market values and "willingness to pay" more troublesome than in areas that touch on social or moral values and issues, such as the valuation of human health or life, which are things that obviously escape pricing. Because of its sterile and clinical treatment of information, the CBA does not deal with moral issues well or, at minimum, it deals with them rather poorly and perhaps should not be applied to them at all. When moral considerations of right and wrong need to be included in an analysis, the CBA is probably not an appropriate tool. In these cases, discussions and debate of the issues should probably replace the traditional CBA, so that issues of right and wrong, as well as legal and policy contexts can be considered. One author says that the inconsistent valuation of life/health is not an issue because CBA values policies or specific risks and not life/health itself. The policy is priced, not the consequences of the policy. However, this approach still excludes the role of discussion and debate in arriving at a decision, which may render it a less desirable option.

Marketed resources can be valued directly from current prices, from values inferred from surrogates (such as similar commodities or services) or estimated from market trends or surveys (e.g., "willingness to pay"). Some examples of values used in CBA are:

- Marketed resources costs: values of the market-priced resources or inputs used in implementing the activity, which include startup capital, operating expenses
- Marketed output benefits: these benefits represent the increased value of physical goods and services that are created by the activity
- Marketed output costs: negative consequences for some individuals resulting from the activity

Cost-Effectiveness vs. Cost-Benefit Analysis

In cases where it is difficult to assign a value to a hard-to-measure nonmarket effect, the analyst can use a cost-effectiveness analysis (CEA). Similar to the CBA, the CEA values policy consequences in monetary terms, with the difference being that at least one policy consequence is not valued but instead quantified in physical units. For example, when evaluating how a local community food garden may be improving local residents' diet, the analyst could estimate how individuals' diets have improved and what that would mean (statistically) in terms of longer life spans. By quantifying the number of life years gained (a physical unit) because of better nutrition instead of assigning a value to the life, the analysis avoids the moral conflict of valuing life. However, this approach also provides less information to the decision maker. The CEA can be valuable in some situations, particularly when you need to value the costs and quantify the effect of a single policy or food growing operation, and assess a number of policies and/or policy levels to identify which minimizes the costs of achieving a given effect or maximizes the effectiveness of a given activity. When comparing the CEA with the CBA, the CBA provides greater information by valuing society's preferences concerning all policy consequences that generate benefits or costs. The CEA does not measure society's preferences regarding the activity. The CEA can be advantageous in avoiding explicit valuation when a particular consequence is controversial (such as valuing a human life or a moral issue).

The CBA and Urban Agriculture

Applying a CBA to the urban agricultural setting can be straightforward, but since agricultural operations tend to be varied in terms of scale, growing methods and purposes, the CBA must be tailored to meet the needs of the particular operation. Certainly, in the case of urban agriculture, one size does not fit all!

Time Valuation

For commercial farms, the expense of labor (time investment) can be monetarily estimated from paystubs and receipts; these values can be directly entered into the CBA. However, for non-commercial operations (e.g., subsistence farms, community gardens and residential food gardens), there is no direct way to calculate the value of expended labor. Instead, other valuation methods must be used and these methods can yield significantly different results. This is because a housewife who works the home vegetable garden may value her time differently than a busy career woman who tends a plot in her local community garden. Below are some possible ways that labor may be valued, for the purpose of an urban agricultural CBA, by non-commercial food growing operations:

- Survey laborers about how much their time would be worth if they were working their regular jobs. This approach assumes that these workers provide their time to an employer for a fixed and agreed upon rate, and that they would also agree to receiving the same payment for working the food growing operation
- Survey laborers about how much the feel their work time is worth
- Survey employers in the area to determine wages that are being paid for a comparable job
- Survey local staffing services to determine wages that are being paid for a comparable laborer
- Determine how much that the laborers' work produces (in monetary value) for the food growing operation and pro-rate that value to a wage

In the case of the latter bullet (above), the labor is not valued for what it costs, but for what it produces. This perspective is most important to the non-commercial food grower who may not particularly care about how much time they invest (within reason) because they are most interested in producing food for their consumption. For them, the quality of the food and the intangible benefits they receive from the activity of growing food is worth their time investment.

Growing Methods and Materials

Many urban residents who grow food may decide to use methods or materials that are relatively expensive- vertical gardening, container gardening, hydroponics, etc. Some of the reasons for these expensive investments may lie beyond the practical or efficient. Homeowners or community gardens may invest in materials or equipment that are decorative or fit into the garden's functional objectives. Some are acquired for teaching or demonstration purposes; other serves multiple purposes. Because the CBA is based on examining the economic efficiency of a food growing operation, the purchase of materials that are not economically efficient runs counter to the purpose of the analysis. This must be noted when examining the results from the CBA.

Purpose of the Food Growing Operation

The objectives of the farmer (or food grower) will determine the values and importance of factors that influence the food growing operation. For example, the maximization of profit over a long period of time (e.g., several decades) may involve a different set of choices and operational procedures than would be made if the food grower were interested in maximizing profits for each of those years (Debertin 2012). Long-term profit maximization may require making larger initial expenditures on durable inputs (e.g., infrastructure and machinery).

Non-commercial urban agriculture operations have different sets of objectives and purposes for existence. Because of this, each will endow the CBA with a different focus and different set of factors, benefits, costs and impacts. To complicate matters, each of these operations may have a different set of intangible, and invaluable, benefits that may be regarded with relatively high importance. This variety of purpose and valuation of intangible benefits makes comparisons between two non-commercial food-growing operations tenuous. This also emphasizes the necessity for the analyst to identify the intangible benefits of a food growing operation.

Water

Water may be expressed as a monetary value, but that too varies by location and context. For example, placing a value on rainwater for irrigation can be difficult. Captured and stored rainwater is very inexpensive once the cost of the capture and storage equipment has been covered. A self-supply water system (groundwater well or pump connected to a surface water body) can be operated with a minimal cost, usually electricity and pump expenses. These expenses can be reasonably estimated. However, in some regions, water withdrawals from groundwater and surface sources contribute to the unsustainable use of water resources. For example, in coastal areas with shallow aquifers, the combined effect of many coastal water supply wells can contribute to the landward movement of saltwater within the freshwater aquifer. Once coastal freshwater wells are contaminated with saltwater, a new well must be drilled at a substantial expense. Urban agricultural operations that draw from at-risk or stressed aquifers contribute to an environmental problem that is important, but can be difficult to monetarily quantify.

Urban food growing sites that are installed as an alternative to more water-expensive land use types (e.g., irrigated lawn) can reduce the risk of unsustainable use of or saltwater intrusion into water resources. These operations may generate a net benefit for the community, which may be monetarily valued according to how that operation reduces the risk of a costly event. This reduction in risk may be expressed as the amount of expenditure that has been avoided by risk reduction, pro-rated by water resource users.

Food Valuation

For commercial food producers, the value of what has been produced usually arises from sales. This monetary value can be determined from cash receipts. In the CBA, this value is listed as a profit. For non-commercial food producers, the value of what has been produced arises from the amount of money that was not spent on produce. This value can be determined from what the grower would have spent on the food that was grown. For these growers, this value is listed as a profit in the CBA. Determining this value can be done by referencing prices from local retail stores for similar products. However, this can become complicated.

Retail stores often charge different prices for the same produce, depending on the store's overhead expenses. Some stores offer special prices on some items to lure shoppers; these prices are usually at or just slightly above the wholesale cost to make them more attractive to shoppers and do not adequately reflect the market price. Many stores may offer conventionally grown and organically grown options of the same type of fruit and vegetable, and the prices for these can be very different. In other cases, the variety of a particular fruit or vegetable offered for sale at the retail outlet may not match that which was grown in the garden. When gathering prices for the purpose of determining the value of what was grown, the best price fit may not always be available and some degree of estimation and extrapolation may be necessary.

Energy and Greenhouse Gas Emissions

The use of energy by commercial and non-commercial food growing operations may be obtained from meters or receipts. This would be listed in the CBA as an expense in terms of expenditures and there would be monetized impacts to the environmental associated with the consumption of non-renewable energy. However, in some cases an urban food growing operation may wish to calculate the savings to energy resources that occurred because food was grown and consumed locally, rather than being involved with long-distance shipping. Estimates of these values can be done using a method that is similar to that described for estimating the monetary value of what was produced-by consulting with produce suppliers or retail outlets to determine the distance that the food produced by the food growing operation would have travelled had it been purchased. In contrast to the commercial operation, the non-commercial producer would enter the amount of monetized value of not using an unsustainable resource and not producing its associated CO_2 emissions. These calculations are not straightforward, but are further discussed in Chapters 12 and 14.

Chapter 11. Quantifying Food Sustainability

What is Food Sustainability?

Before wading into the topic of food sustainability, it is necessary to first define what food is and where it comes from (the concept of sustainability was previously discussed in Chapter 4). The term "food" globally refers to a material that is consumed for sustenance and nutrition. But, the specific plants and plant products that have been identified as food are not universal, immutable facts; indeed, the definitions of what constitutes food varies from region to region and from culture to culture. Recall the example from a previous chapter about sweet potatoes; the sweet potato tuber is the only plant part eaten in some cultures but in other cultures, the leaves and stems of the sweet potato vine are the most desired edible part of the plant. A small farmer may grow sweet potatoes, but the culinary preferences and heritage of the local residents will determine which part of the plant is considered to be food. To complicate matters, a particular part of a plant may be used as food by a certain cultural group during one point in time, but not another. For example, German-Americans once consumed both the leaves and the bulbous stem of the kohlrabi plant, but today this same ethnic group only consumes the bulbous stem. When considering or calculating food sustainability, it is essential to have a definition for "food" that can be universally applied. For the purposes of this book and the upcoming analyses, food is defined as the amount of material that is available to be consumed for sustenance or nutrition regardless if it is utilized by a cultural group or not.

This more liberal definition of food has implications for the calculation of productivity and food sustainability (these topics will be explored in later sections). However, to use a more restrictive definition, one that is based on the culinary preferences of a certain group of people, confines the relevance of an analysis to a specific group at a specific time. The benefit of using the more inclusive definition is that it can account for the total amount of food that could be available. This amount can then be divided into subsets that reflect the local residents' food preferences by excluding those products (from analysis) that are not normally consumed.

Another difficulty arises when one seeks to calculate food production. Investigators who seek to measure the amount of food produced in an area or that is produced within a certain distance from a city (the foodshed) often rely on published data from commercial food operations. However, this is only a subset of the total amount of food that is grown or available to humans. For example:

- Farmers may not harvest all of the food products that are grown; sometimes not all of the crops are harvested for market
- Food produced by non-commercial operations, in residential settings or in community gardens are not normally quantified and reported
- Food can be obtained from sources other than agriculture, such as hunting and gathering wild plant products, which can be an important food source for people living in some (usually rural) areas

The last bullet above mentions an interesting and important aspect of the food system- the production of human food by natural ecosystems. The amount of food that enters or bypasses the food system from non-agricultural sources is poorly accounted for and is largely ignored in foodshed calculations, which are biased towards commercial agricultural production. It is easy to overlook the fact that, until the 20th century, a substantial number of Earth's inhabitants relied on wild foods as an essential part of their diets, and a number of long-lived cultures still do. The production of wild food (and its contribution to human health and nutrition) lies beyond the scope of this book and will not be included in further discussions. This fact is not meant to diminish its importance.

At its core, the goal of sustainable agriculture is to support the three sectors of sustainability (Figure 1). Economic sustainability is important because food-growing operations whose costs exceed gross benefits (sometimes expressed as monetary value) will not survive over the long term. In the same context, a food growing operation that does not contribute to food security or well being at the individual or societal level will not be successful in the long-term. Although the relative importance of each of these two issues will vary from place to place, they play a part in the longevity of any food growing operation.

Lastly, the environmental sector of sustainability demands that food must be produced in ways that do not damage or destroy the resources upon which the practice depends. As explained in an earlier section, if a food growing practice consumes a limited resource, or consumes a resource faster than it is regenerated, the practice is unsustainable…and it is the nature of unsustainable practices to become obsolete. By extension of this concept, the food production must be done in a manner that does not damage or destroy the environment, in either directly (e.g., the farm field) or indirectly (in distant locations). But, food sustainability includes considerations that lie beyond the farm field.

Urban residents are on life support with respect to their food supply. This is because they do not grow enough food for themselves or the area they live in does not produce enough food to feed the local residents. City residents depend on someone else to grow and transport their food to the retail outlets, where it is purchased. This places urban residents in a precarious (and unsustainable) situation for several reasons. First, these residents have lost some degree of control of their food supply and choices. Second, if someone cannot pay what the market demands for the food product or if food is not available, that person will not eat (unless supported by food assistance programs). Third, for urban residents that do not grow food, it is a market commodity that they must constantly spend money on. Because of income inequality, it is inevitable that there will be food access inequality. Besides these impacts to urban residents, the production methods of large-scale commercial food growing operations and long-distance transport of food into cities creates pollution, impacts natural habitats and consumes non-renewable resources. These reasons, among others, demonstrate why urban areas are food unsustainable. They also underscore why local and sustainably produced food is essential for food sustainability. When a city's food supply comes from local sources, instead of distant sources, there is can be a higher degree of food security (Burton 2012). *During the remainder of this book, the concept of sustainably produced food will include the assumption that the food was produced locally using sustainable growing methods.*

The concept of sustainable agriculture is predominantly focused on the practice of food growing. Once the food leaves the realm of the farm, it enters a new domain where different laws, policies and valuation structures influence where the product ultimately goes to and where it is consumed. The complete set of elements and relationships between the place where the food is produced, food products, economy and community are referred to as the food system. Although the agricultural practices used to produce products for the food system may be run in a sustainable manner, this does not guarantee that the food system is sustainable.

A sustainable food system includes a broad set of considerations that include:
- Crop production (planting, growing, harvesting)
- Processing, packaging, transporting
- Marketing
- Food access
- Equity concerns
- Consumption
- Disposal

Factors that can influence a food system include social, economic, political and environmental considerations, and several models have been forwarded for food system analysis. What needs to be done to achieve agricultural sustainability has become clearer over the past decades, but the task of defining how to make a food system sustainable is much more complex.

Besides the terms "sustainable agriculture" and "sustainable food system", a new term has arisen the past decades- food sustainability. Although there is a broad consensus that it is a good thing to attain, there appears to be little consensus about what it actually means. An examination of Internet searches on the term yielded a number of related topics, but failed to define a specific, unique concept for "food sustainability." Without exception, the term is used without an explicit definition. Some examples of how the term "food sustainability" has been used include:

- Food sustainability encompasses many interpretations and aspects such as ecology, economy and society; in this example, the use of the term food sustainability is equated with sustainable food system. (Aiking, Harry & Joop De Boer 2004, Garnett 2013)
- Food sustainability as a function of food security and nutrition (Lang, Tim & Barling 2012, Jones *et al.* 2012, Premanandh 2011, Wahlqvist 2004)
- Food sustainability defined as the ability to meet food needs in a sustainable manner (Helms 2004).

A common thread that is found throughout these definitions and usages of the term food sustainability is the relationship between sustainable food production and the ability to *adequately* meet the food needs of a population. The qualifying term "adequately" would require that the food be produced in the quality, quantity and types to meet (at least) minimum human food consumption requirements. Otherwise, community health and welfare would be impaired unless outside sources of food could be secured and imported. The different uses of the term food sustainability also implied an association between food sustainability and food security, the latter of which addresses not just the supply of food but also people's access to it.

In this chapter, a more precise meaning of food sustainability is defined and applied- one that is quantifiable and meaningful to efforts to understand how close a community is to achieving food sustainability. To summarize the discussion in the previous paragraph, food sustainability is achieved when: (a) sustainably produced food (b) is available in the quantity and types needed to provide a healthy diet (c) for all of the residents of an area. Note that (a) addresses economic and environmental sustainability, (b) and (c) addresses social sustainability. Accordingly, a definition for food sustainability can be described in both qualitative and quantitative ways, through the calculation of metrics:

- Food sustainability can be conceptually defined as the ability to meet the dietary needs of a population through locally and sustainability produced food.
- Food sustainability can be quantitatively measured as the difference between the amount of food that is or can be (locally and sustainably) produced and the amount of food needed to provide for a healthy diet for a population; from hereon this calculation will be referred to as the Food Sustainability Measure (FS_M).
- The Food Sustainability Index (FS_I) is the ratio between the amount of food that is or can be (locally and sustainably) produced and the amount of food needed to provide for a healthy diet for a population; when the FS_I value is equal to or greater than 1, the population has achieved food sustainability.

When suitable data is available, calculation of the FS_M and the FS_I gives an objective measure of the food sustainability condition of an area and can provide a way to compare the purported benefits of existing and proposed urban agriculture projects. The above definitions are flexible with respect to the size of the food producing area (the foodshed, which is described in detail in the next section) or the population they are applied to. This scaling flexibility allows the quantification of the food sustainability metrics to be quantified for areas ranging from local (e.g. neighborhood) to national levels.

The definitions of food sustainability presented in the last two bullets above (FS_M and FS_I) can be expressed as equations. To calculate the FS_M, subtract the amount of food needed (food demand or FD) for the population of an area (FD_P) from the historic food production (FP_H) in the foodshed, as shown in the following equation:

$$FS_M = \text{Foodshed } FP_H - \text{Population } FD_P$$

The result will be a number that is positive, negative or approximating zero. If the resulting number is negative and large, then there is a great imbalance between the food demand and the amount of food produced in the area of interest. The FS_M indicates how much more sustainably grown food must be produced in order to achieve food sustainability. If the result is a number that is close to zero, then the area is considered to be food sustainable. If the result is a number that is positive, more food is produced in the area than residents need to consume. The area is then both food sustainable and generating a surplus.

Similarly, the FS_M can be calculated from potential food production:

$$FS_M = \text{Foodshed } FP_P - \text{Population } FD_P$$

The results from the FS_M calculation from potential food production are interpreted in the same way as the calculation of FS_M from historic production data. However the result indicates a hypothetical condition of some kind, such as a proposed scenario. This type of analysis is valuable to demonstrate how a particular proposed activity could contribute to an area's food sustainability.

The FS_I is a single number that indicates the food sustainability condition of an area of interest and is useful to draw comparisons between different areas. When the FS_I is calculated using historic food production values, the resulting FS_I reflects the current (or past) condition of the area:

$$FS_I = \text{Foodshed } FP_H \div \text{ Population } FD_P$$

When the calculated FS_I value is less than 1, the population (FD_P) is not food sustainable. When the FS_I is close to or greater than 1, the population's food needs are being met sustainably. The FS_I may also be calculated using the potential food production value; the resulting index value will reflect a hypothetical condition. This is valuable for comparing between different proposed food production scenarios and understanding which would yield the greatest benefit:

$$FS_I = \text{Foodshed } FP_P \div \text{ Population } FD_P$$

Calculation of the FS_M or FS_I requires that the following factors are known or estimated: the food demand of the residents in the area of interest and the quantity of food that is or could be sustainably produced. More about these factors, along with how to calculate the food sustainability metric, are described below.

Quantifying local environmental sustainability through metrics is not a new concept. Gosh, Vale & Vale (2007) published an analysis of the potential local environmental sustainability of five residential urban forms from case studies in Auckland, New Zealand using five main ecological footprint parameters: domestic energy, transportation, vegetation cover, food and waste. This quantitative study formulated a comprehensive methodology for measuring comparative sustainability performances of different development forms.

In a separate study, Gosh, Vale & Vale (2008) analyzed local vegetable productivities associated with available productive land areas as home gardens in low, medium and high-density residential urban forms using aerial photointerpretation, GIS and mathematical methods. The food production potential of a household in five different residential urban form case studies from the Auckland Region, New Zealand were compared to determine the most "food efficient" urban form.

Quantifying Food Production

Quantifying food production, whether actual or potential, is a rather complex exercise. At best, imprecise estimates can be made. However, the fact that exact numbers cannot be obtained should not discourage the analyst from doing so. In fact, estimates that are derived from reliable sources that are presented with the appropriate caveats and with the known bracket of uncertainty are highly useful in understanding the magnitude of food production from an operation or spatial unit. Uncertainty analysis can be conducted to provide understanding about the level of imprecision carried by a food production or food demand estimate.

When examining the question of food sustainability, one must begin with the human diet and work from that platform in order to define how much food must be produced to meet the food needs of a population. From a social sustainability perspective, a food system that does not meet the food needs of the population is inadequate. The types and quantity of food must, at minimum, match the requirements for a healthy diet. If it does not, then that food system does not adequately contribute to long-term human health. Much of the current commercial agricultural system is focused on producing crops that have the highest potential for making a profit or which have the greatest market demand. Because of this, the products from the commercial agricultural system reflect market trends and personal food preferences, rather than human dietary needs. These influencing factors cause the commercial food market system to be inefficient with respect to meeting human dietary needs and, as such, ineffective in achieving food sustainability.

The Foodshed

A foodshed represents the geographic area where food moves from its point of origin to its market or consumer (Hedden 1929, Getz 1991). The foodshed contains all of the activities that produce food for a target population (the population that is considered in the food sustainability analysis). In earlier times in human history, the foodshed was located relatively close to where residents lived. Throughout the 20th century, food transport expanded step-wise from local to continental to international to global scales. Today, a strict interpretation of a foodshed, from the perspective of an urban resident in a developed country, would reach into most continents around the world. When defining the boundaries of a foodshed, this unwieldy reality can be brought down to manageable extents by narrowing the question being asked of the foodshed analysis.

It is important to recognize that agricultural operations are not the sole way that food is produced. As mentioned earlier in this chapter, the total amount of food available may include non-commercial and under-counted sources such as home food gardens, community gardens and natural areas that are utilized for hunting and foraging. Quantifying the amount of food produced by commercial operations can be relatively easy, as compared to the task of quantifying the amount produced by urban agriculture or natural ecosystems. In urban areas, where most residents are entirely dependent on the commercial food system, the foodshed may be more easily determined from retail and wholesale purchasing records. But in rural areas, residents may be more dependent on food obtained by hunting, foraging, home food growing and neighborhood fruit and vegetable stands; quantifying productively from these activities is much more difficult to determine.

Conceptually, the foodshed can be viewed in two distinct ways: (1) the sum of all agricultural crops produced by food growing operations in the foodshed, or (2) the sum of agricultural crops produced by food growing operations in the foodshed that contribute to a healthy human diet. These calculations are expressed as units produced through time. The conventional expression of agricultural productivity is the volume or weight of food produced over a growing season or year. That convention will be used throughout the rest of this book, unless otherwise stated.

Peters *et al.* (2009a) state that "Providing a wholesome and adequate food supply is the most basic tenet of agricultural sustainability." When the goal is to achieve food sustainability or to calculate a food sustainability metric, food productivity measurements must be considered from the perspective of what entails a healthy diet; therefore, food productivity measures must be filtered through the sieve of dietary guidelines for a healthy diet. Agricultural products that are outside of these guidelines, although profitable or marketable, are secondary and do not contribute towards achieving the goal of food sustainability. So, when quantifying food production for the purpose of calculating a food sustainability metric, it is important to keep in mind that production must be specifically focused on crops that contribute to a healthy diet. This criterion will be referred to as the *foodshed efficiency*. The foodshed efficiency, which can be measured by the FS_M, describes how closely agricultural production in the local foodshed matches the food needs (demand) of the population within the foodshed. Regions that have the highest foodshed efficiencies are those whose agricultural operations produce a wide range of crops in the relative abundances to meet the guidelines for a healthy diet. More on dietary requirements and calculation of a foodshed efficiency will be discussed in a later section.

In general, there is no consensus in the scientific literature for what "local" means with respect to the distance between the food grower and the consumer (Martinez 2010). The spatial extent of the foodshed will vary, depending on the objective of the analysis. For example, if one wishes to examine the magnitude of the deficit or surplus of food that is being produced by local agriculture, then the extent of the foodshed would be delimited according to the definition of "local" agriculture that is applied. The area may extend well beyond the cities that would be fed from it. Some sources have defined local agriculture in the following ways:

- U.S. Congress' 2008 Food, Conservation and Energy Act (2008 Farm Act) defined a "locally or regionally produced agricultural food product" as having travelled less the 645 km (400 miles) from its origin or within the state it was produced
- San Francisco Foodshed Assessment (b08)- 160 km (100 miles) from the Golden Gate
- New York State Foodshed Assessment (Peters *et al.* 2009b) used the state borders as the source of locally-produced food

In another example, one may want to determine how much food could be produced within a metropolitan area's greenspace. In this case, the foodshed would correspond to the metropolitan area's borders. In a third example, one may want to know how much area around the target population would be needed to meet their food needs. Each of these applications of the concept of a foodshed is useful to answer different questions. Before embarking on an analysis of food sustainability, the boundaries of the foodshed must first be established.

To summarize:

- In order to calculate or estimate food production from the foodshed, it is necessary to know the amount of food that is (or could) be produced from the food producing operations
- The spatial extent of the foodshed must be defined; in some cases, what is meant by "local" food must be set by the analyst
- The food produced must contribute to a healthy diet, both in variety and quantity

Food production can be calculated from survey data, censuses or studies, or may be inferred from proxy crops. These data may also be referred to as *historic production data* because they were collected and compiled after-the-fact. There are times when food production over an area or foodshed may need to be estimated. This is particularly true when one wishes to predict future production values or extrapolate production data across areas that lie beyond the food-growing field. Using (proxy or actual) food production data for predictive or extrapolation purposes will be referred to *potential food production*. How historic and potential food production data can be enumerated is described later in this chapter.

The "Healthy" Diet

Before the historic or potential food production from a watershed can be calculated, it is first necessary to constrain the definition of food productivity to those agricultural products that contribute to a healthy human diet. One widely accepted source of the food types and quantities that should be consumed in order to have a healthy diet is the United States Department of Agriculture's (UDSA's) Choose My Plate guidelines (USDA 2016). These guidelines describe the recommended number of servings that should be consumed by men or women in different age ranges for six main food groups: protein foods, grains, oils, dairy products, fruits and vegetables. Some of these food groups are further divided into subgroups; for example vegetables are categorized into five distinct subgroups: dark green, red/orange, starchy, beans/peas and other vegetables. The amount of any food group an individual needs to eat depends on that person's age, sex, and level of physical activity. Recommended total weekly amounts of fruits and each of five subgroups of vegetables that should be eaten are shown in Table 12. To demonstrate how the foodshed production and food sustainability metrics can be calculated, this chapter will only cover fruit and vegetable groups. It is recognized that the other food types (e.g., protein foods, grains and oils) are also necessary for a healthy diet. They may be included in an analysis of food sustainability using the same methods described for fruits and vegetables below.

The guidelines presented in Table 12 place less emphasis on the actual varieties of crops that are grown. Instead, crops types that are nutritionally similar (e.g., dark green vegetables such as spinach, kale, endive, mixed greens, are bok choy) are combined into the same subgroup. Table 16 lists some of the vegetables and that are included in the five subgroups of vegetables shown in Table 12.

Using the Choose My Plate guidelines (Table 12), foodshed productivity can be expressed as the total annual amount of fruit and vegetables produced within the area of interest each growing season (year). Productivity data may come from one source or may be combined from several sources. It is unlikely that all of the needed information can be obtained for some areas and estimates may be necessary to complete the data sets. Some of the possible sources of the agricultural production data are described below.

Quantifying Historic Food Production (FP$_H$)

Quantifying the FP$_H$ of a foodshed is straightforward and can be achieved through the following steps:

1. Delineate the extent of the foodshed that will be part of the analysis
2. Identify food producing activities within the foodshed
   A. Identify types of food producing activities in the foodshed

Table 16. Vegetable varieties within Choose My Plate vegetable subgroup.

Dark Green Vegetables	Red/Orange Vegetables	Starchy Vegetables	Beans and Peas	Other Vegetables
Arugula	Acorn squash	Cassava	Black beans	Artichokes
Bok choy	Butternut squash	Corn	Black-eyed	Asparagus
Broccoli	Carrot juice	Green bananas	peas (dry)	Avocado
Collard greens	Carrots	Green lima beans	Chickpeas	Beets
Dk-green leafy	Chili peppers	Green peas	Kidney beans	Brussels sprouts
lettuce	Pattypan squash	Jicama	Lentils	Cabbage
Endive	Pumpkin	Plantains	Navy beans	Cauliflower
Escarole	Red bell peppers	Potatoes	Pinto beans	Celery
Kale	Sweet potatoes	Taro	Soy beans	Cucumbers
Leeks	Tomato juice	Water chestnuts	Split peas	Eggplant
Mesclun	Tomatoes		White beans	Garlic
Mixed greens	100% vegetable juice			Green beans
Radicchio	Yams			Green peppers
Romaine lettuce				Iceberg lettuce
Spinach				Mushrooms
Swiss chard				Okra
Turnip greens				Onions, Scallions
Watercress				Radishes
				Red cabbage
				Snow peas
				Sprouts
				Tomatillos
				Turnips
				Wax beans
				Summer squash

       i. Commercial operations

      ii. Non-commercial operations

     iii. Other food producing activities

3. Calculate the annual amount of food produced within the foodshed

    A. Summarize the known amount of food produced by each food producing activity from:

        i. Reports

       ii. Surveys

      iii. Census records

      iv. Other resources

    B. Estimate unknown food production by each food producing activity by:

        i. Calculating the total area of each food producing activity whose food production quantity is unknown by:

        a. Consulting land use maps

        b. Conducting aerial photo interpretation

        c. Conducting on-the ground surveys and groundtruthing

        d. Other resources

    ii. Determining historic productivity from each type of food producing activity from other similar regions or estimating from published sources

Food production data can come from a number of sources and can be of different degrees of quality and quantity. The acquisition of production data of sufficient granularity can be a fact that constrains the analysis and limits the accuracy of the calculations. Researching the weather conditions over the growing seasons that the data were collected is recommended to understand if these measurements are from a more normal climatic period. Data collected during drought years or prolonged heat waves may understate productivity that may be expected under normal or more optimal climatic conditions.

Productivity values can be proportionally scaled to infer how much of a certain type of produce may be expected from a production area. For example, if an agricultural survey reported that 5 tons of green vegetables were produced from farms encompassing 10 acres, then one can infer that 0.5 tons of green vegetables may reasonably be grown on a field that is 1 acre in extent. There are limits to how far one can stretch these inferences, but one order of magnitude may be reasonable. Where productivity data is not available from a site, they may be inferred from production data collected elsewhere within the same region.

Historic production data can come from many sources and it is essential to understand how the data were collected and the relative degree of precision that the reported numbers carry. Be aware that reported agricultural production numbers do not infer if the agricultural products are being sold and if so, to whom or to where. The most-studied agricultural crops are those that are commercially important and have a high profit potential; these may not be the same crop types that one wishes to include in their analysis. Below are some possible sources of historic agricultural production data.

Agricultural Census Data and Surveys

The United States Department of Agriculture conducts an agricultural census every ten years, on years that end with a "5". These data are available through the Internet and may be useful to calculate the gross productivity of a certain agricultural crop in a region of interest (https://www.agcensus.usda.gov/index.php). Data are usually reported by county and can include the number of acres in production (harvested) and some kind of production value (weight, units, etc.). By calculating the average productivity per unit area, the food production from a specific area can be determined.

Another approach that can be used to quantify food productivity is to survey local growers. Many growers keep records of how much of an area is planted with crops, how much was harvested and the cash receipts received for those crops. Information such as this is valuable because it may better reflect local growing conditions than data obtained from other regions. Be aware that this is confidential information for most commercial operations and that some growers will not be willing to yield the data without a confidentiality agreement.

Crop Production Studies

Crop yields for a variety of fruit and vegetable varieties are usually available from agricultural studies that were specifically designed to measure the productivity of different crop types. Internet websites, agricultural databases, horticultural publications and state agricultural universities often make these data available to the public at no cost. These data can be valuable, but because the studies were conducted under more ideal conditions than urban food growers usually find themselves in, the data may overstate or understate expected productivity.

Commercial Agricultural Crop Data

Agricultural productivity data has been published for many commercially important crops. However, there are some limitations on using these data, including:

- Crop production values are derived from agricultural fields rather than urban plots, which use different planting, harvest and maintenance methods
- Agricultural studies often grow crops in a monoculture, such as is typically found in commercial agricultural operations; these conditions are not typical for small-scale urban agricultural operations
- These studies may focus on agricultural crops that are the most economically important, a consideration that is seldom important for some types of urban agriculture (e.g. home food gardens, community gardens, etc.)
- These studies often plant hybridized or genetically modified varieties that are usually not grown in urban settings

Proxy Data

Agricultural production data from one type or variety of crop may be used as a proxy for another, provided there is sufficient justification for doing so. For example, if there is good production data for a variety of leaf lettuce, then one may use that as a proxy for other similar varieties of leaf lettuces- so long as the growth habits are comparable. However one should not substitute production data from dissimilar varieties, for example, the relatively small acorn squash for the relatively large Hubbard squash, unless there are production studies that indicate total yields from both are similar over a growing season. When there is a paucity of data from the foodshed of interest, data may be used from regions that are climatically, topographically and geologically similar to the foodshed may be a reasonable proxy.

Other Crop Data

The most desirable data to use to calculate potential food production is from studies conducted in the region of interest and that measure productivity from a plot that yields a variety of food crops that are more closely in line with human food need. This type of food production is most commonly practiced in urban agriculture (home food gardens, community gardens and urban/peri-urban farms). An example of this type of data will be presented in Chapter 13.

Quantifying Potential Food Production (FP$_p$)

Potential food production (FP$_p$) is the amount of food that *could* be grown within a specified geographical area under certain scenarios that do not currently exist and can be applied at different spatial scales according the question being asked. The FP$_p$ is valuable for conducting "What-If" scenarios to determine:

- How much food could be produced from existing production sites
- How closely existing food production sites can meet the current or future food demand of a population
- How closely a proposed food production scenario would meet the food demand of a population (e.g., the implementation of a new food production operation)
- The relative benefits of different food production scenarios.

The latter exercise is an important analysis for communities that want to look at different options for increasing local food production and to identify which production scenario would most closely achieve food sustainability targets.

The FP$_p$ can be calculated for the area where the food is produced and where the consumers reside (e.g., within a city); because the food is both produced and consumed locally, the food production is considered to be internal to the area. In some cases, the analyst wishes to calculate the food that can be produced within a certain distance from where urban residents live (e.g., within a foodshed). In this situation, food may be produced within the same region where the residents who consume the produce live, but the food production may not occur within the same city where the consumers live. This is an example of external food production.

The FP_p is calculated from two values: the amount of area available for food production and the expected or likely productivity per unit area. To begin, the analyst must define the amount of area available for food production, which may not be the same area as the foodshed. If an analysis is being conducted to determine the potential production from a single or small collection of urban food growing operations, then the spatial extent of the analysis cannot be considered to be a foodshed. In this case, the food production area is the growing space available to the operation (or the combined space of the collection of operations). For small operations, the extent of available growing space can be physically measured. However, for landscape-level analyses that seek to understand how much food can be produced within a region (e.g., a metropolitan area), the analyst may wish to view the total area included in the analysis as a foodshed. For larger outdoor areas, aerial photography, GIS and interpolation methods may be needed to define the extent of land available for food growing.

Several methods exist for quantifying land cover types within urbanized areas, some of which can be used to determine the extent of potential food production areas. Histogram stretch, supervised classification and unsupervised classification techniques in GIS are methods that utilize digital image processing to discern open land. Because these are automated processes, verification procedures will be necessary to validate results. Another approach to quantifying available land in urban areas is to manually identify and measure areas of interest on aerial images. For smaller areas, the entire area of interest may be examined. However for larger areas, a randomized sample of the area of interest (using a statistically appropriate number of sample points) can yield a reasonable estimate of the amount of growing space that may be available. When conducting a landscape-level analysis, it is important to remember that not all urban food growing sites identified using aerial survey methods may be usable since issues such as access, permission to use the property, ownership issues, and land suitability may restrict food growing potential.

Historic crop study data or crop production data can be used to infer likely future or expected crop production (see previous section). These data must be represented in quantities that are relevant to the healthy human diet (Table 12) if the analyst seeks to report the results in the context of sustainability.

Quantifying Potential Food Demand (FD$_P$)

The Potential Food Demand (FD_p) is the amount of food that is needed to provide a healthy diet to the people that are living in a defined area. For most studies that examine the food needs of a city, sales statistics (at retail or wholesale levels) of various food products are compiled and used as the baseline for determining whether or not a city's food needs could be met through local food production. These sales statistics may represent the actual consumption of food products, if it is assumed that a negligible portion is lost as waste. However marketing, advertising, habit, cultural food preferences, food availability, food affordability and other factors influence consumer-purchasing choices. It is well known that there is disconnect between what many people purchase for food and what constitutes a healthy diet. To make this issue more complex, food preferences change through time…what the majority of people eat today is quite different from what people normally ate 50 years ago. Although food choices change through time, the biological need for a balanced, healthy diet remains unchanged. Because of this, the food needed to sustainably provide a healthy diet must be based on a set of fixed dietary standards rather than actual consumption.

Two values are needed to calculate the FD_p: the number of people living in an area and the amount of food they need to consume to remain healthy. The first value, the number of people living in an area, can be obtained from census records or other population estimation studies. In the United States, census data and corresponding demographic data for statistical areas (cities, counties, metropolitan areas, etc.) can be obtained from the United States Census Bureau (www.census.gov) and other websites. The amount of food needed by the population of an area can be calculated from the USDA Choose My Plate guidelines (Table 12). Age demographic categories, as defined in the Choose My Plate guidelines and the US Census Data, differ slightly. To rectify these differences, the Choose My Plate age categories were adjusted to provide compatibility with the US Census age demographic data; these adjusted values are shown in Table 17 . By multiplying the number of people within the different demographic groups (age, sex) obtained from census data by the amount of food needed to be consumed by each group (Table 17), the total amount of food needed to provide a healthy diet for that population can be estimated. Note that the food demand is a fixed quantity that is based on population.

Table 17. Modified Choose My Plate age categories and recommended weekly consumption (servings) of fruit and vegetables.

Demographic		Vegetables					Fruit
		Dark Green	Red & Orange	Starchy	Beans & Peas	Other	
Sex	Age						
Children (M, F)	<5	0.75	2.75	2.75	0.5	2	7
Girls	5-18	1.5	4.75	4.5	1.25	3.75	10.5
Boys	5-18	1.75	5.75	5.5	1.75	4.5	12.25
Women	18-65	1.5	5.5	5	1.5	4	14
	>65	1.5	4	4	1	3.5	10.5
Men	18-65	2	6	6	2	5	14
	>65	1.5	5.5	5	1.5	4	14

The Food Sustainability Measure (FS$_M$)

The FS$_M$ is a way to quantitatively compare the amount of food that can be or is being produced within a defined area with the amount of food that is needed to feed a local population. The FS$_M$ can be used as part of a foodshed assessment, but it is more constrained than other similar foodshed assessments that have been presented in published reports and scientific articles. Overall, foodshed assessments and have employed several different approaches. Some of these are:

- An approach used by investigators in British Columbia (British Columbia Ministry of Agriculture & Lands 2006), San Francisco (Thompson, Harper & Kraus 2008) and Massachusetts (Holm 2001) quantified the amount of food that was produced within a specified "local" area and compared that to the amount of food consumed (purchased) by residents of that area. These analyses looked dairy, meat, vegetables, fruit, grains and other products.
- Peters *et al.* (2009) developed a GIS-based landscape-level model to analyze the capacity of local agriculture to meet the food needs of New York State population centers. The analysis included an optimization technique to allocate potential production to meet food needs across minimal distances. Models such as this are important because there is a connection between the localness of food systems and sustainability. However, that approach is based on a regional view and assumes that

only urban centers would be the recipients of all of the region's agricultural products.

- Gosh *et al.* (2007) used data from case studies to calculate the potential productivity of urban gardens in Aukland, New Zealand; these results were used to infer self-sufficiency and sustainability parameters.

In a conceptual sense, when a community's food requirements can be met by local foodshed production, this is considered to be a more sustainable condition. Several reasons that this is desirable are:

- The money spent of food is retained within the community (as compared to money spent on food that has been imported to the community from elsewhere), creating a higher level of economic sustainability
- When sufficient volumes of food are produced (locally or within the foodshed) to satisfy residents' needs for a healthy diet, that creates a higher level of food security
- Local food production reduces the amount of pollution and consumption of non-renewable energy sources associated with long-distance transport of food, creating a higher level of environmental sustainability
- If enough food to feed all residents can be produced locally (or within the foodshed) using sustainable growing methods, then that community has a high level of food sustainability

The Food Sustainability Index (FS$_I$)

The FSI is a measure of how closely local (or foodshed) food production is meeting (or is able to meet) local food demand and may be considered to be a measure of the foodshed efficiency; unlike the FS$_M$, the measure is expressed as a relatively small value rather than a set of production and demand numbers. As previously mentioned, FS$_I$ values that range from 0 to less than 1 indicate an unsustainable food condition for the population considered by the analysis. A FS$_I$ value that is approximately 1 or is greater than 1 indicates that the population is in a sustainable food condition with respect to supply and demand (Figure 9).

SUSTAINABILITY CONDITION

| Highly Unsustainable | Unsustainable | Sustainable | Resilient & Sustainable |

FOOD SUPPLY

| INADEQUATE | ADEQUATE | SURPLUS |

| 0.0 | 1.0 | >1.0 |

FOOD SUSTAINABILITY INDEX VALUE

Figure 9. Food sustainability index value relative to the population's sustainability condition and food supply.

Some of the potential uses of the FSI are:

- For estimating how closely local, sustainably produced food is meeting or able to meet the food demand of the residents (foodshed efficiency)
- As a tool to compare the benefits of proposed policies, initiatives or projects to increase local (sustainable) food production (and increase foodshed efficiency)
- For calculating a value that can be easily visualized on a map to view spatial patterns of food sustainability

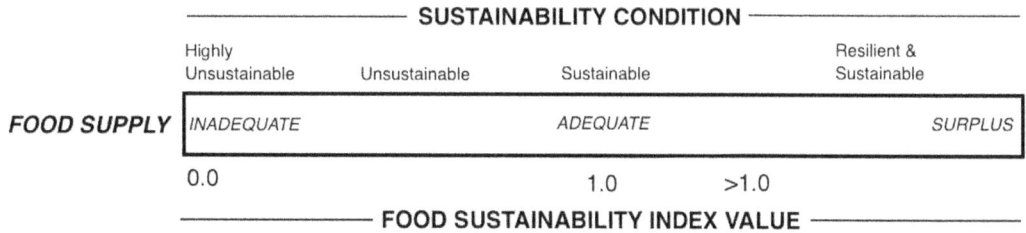

Chapter 12. Residential Food Growing Operation Case Study- Methods

Commercial agricultural systems are fairly well studied and numerous books describe how to operate a farm from the perspective of inputs, outputs, business operations, investment and crop management. In contrast, few studies exist that quantitatively measure the inputs and outputs of urban agricultural systems in detail. Urban food producing systems, particularly non-commercial forms, are perhaps the least studied and least understood forms of agriculture in terms of economic importance, environmental benefits and ecological functions. This is an unfortunate knowledge imbalance that has contributed to the wholesale abandonment of perhaps the most highly productive and efficient form of human food production. This residential food garden study sought to address this deficit by quantifying a suite of benefits from one form of urban agriculture, demonstrating methods for measuring those benefits and adding to the body of agricultural knowledge. The approach used in this study was a *life cycle assessment* of inputs and outputs.

A life cycle assessment is one that examines a full range of known inputs and outputs from an activity over a defined period of time. Some of the inputs and outputs can be reliably measured, some can only be understood in terms of relative magnitude, and some can only be qualitatively described. Although all inputs and outputs may not be accurately quantified, this does not necessitate abandonment of the effort to measure these important values (Adler & Posner 2001, Campbell & Brown 2003, Bernard 2011). In some cases, it is enough to qualitatively or categorically define benefits or costs to demonstrate their importance. More specifics on what was measured and how it was measured are provided below.

During the course of a study, some investigators find that their initial research questions were too broad to be reasonably answered, too narrow to be practical, there were no methods that could adequately measure a factor or that the relative importance of some factors changed as more about them became known. Given these uncertainties, it is in the investigator's best interest to use methods and measures that have some degree of flexibility to allow an adequate data set to be collected and appropriate analysis to be conducted to address study questions. This is normal, particularly for multi-year studies that are at the mercy of environmental conditions. Although this study's goals and objectives remained fixed throughout the data collection period, the depth of investigation into any particular factor was modified as more became known and the limitations of what could reasonably be measured were learned.

Below are presented the methods that were used to calculate sustainability parameters for the residential food growing operation. In some cases, alternative methods for calculating these parameters are given to provide more options to investigators and analysts. This chapter is best used as an example of how one can approach the problem of calculating sustainability parameters and the methods presented should not be taken as the final word on how these parameters should always be calculated. The reason for this is simple. The sustainability sciences are rapidly evolving and a particular data set used in a calculation today may be replaced later by another that is more reliable or precise. That is the nature of science. For example, the estimated fuel efficiency of a fully loaded semi tractor-trailer may be approximately 9 km/gallon in 2015, but the fuel efficiency of these vehicles may be markedly higher by 2025. Because of this, when applying any of the below methods that rely on similar parameter estimates, it is in the investigator's best interest to research and use the best available data before committing to its use. In concert with using the best available data at the time of the analysis, it is also essential to use the data that are most appropriate for the question being addressed.

Urban Food Garden Study's Goals and Objectives

The study's goal was to characterize the economic, environmental, ecological and social aspects (or *effects*) of a residential food growing operation, and to measure costs and benefits to the household. The results from this study were also extrapolated to the community-level to understand the magnitude of benefit that could be realized by many small-scale urban food-producing efforts. The primary research questions from this study were:

- How much can be produced from a small-scale residential food garden?
- Was it worthwhile to grow that food? Specific aspects of this question include:
  - What were the economic benefits of the food growing operation?
  - What were the environmental and ecological benefits of the food growing operation?
  - What were the social and intangible benefits of urban food growing?
- Can enough be grown within this urban setting to significantly contribute to or meet resident's needs?

- How can urban agriculture contribute to community sustainability and specifically to food sustainability?

This suite of questions is more robust than is usually addressed by agricultural studies. This broad reach is necessary to more fully understand and measure the life-cycle costs and benefits of urban agricultural systems. Only when the fuller picture of costs and benefits are understood can meaningful comparisons between different agricultural systems be made with respect to sustainability, productivity and profitability. Further details about the case study's research questions and the methods used to collect data to address them are provided below.

How Much can be Produced?

Agricultural productivity is essential for maintaining a viable operation and has been systematically measured for a very long time. Agricultural research universities and their associated local extension offices can provide farmers with tables of average labor inputs, operating costs and expected harvest for specific crops. Although many of the principles and guidelines published in those books are applicable to commercial urban agriculture, some are not because of the unique nature of growing food within the urban context. One stark difference is the tendency for rural commercial agricultural operations to grow very few crop types. In those cases, quantification of inputs and harvest are relatively straightforward. Urban farms and small-scale urban agricultural systems tend to grow a wider variety of crops, often in a mosaic. The larger menu of crop plants and more variable field conditions that are a part of urban agricultural systems complicate the usage of published single-crop input requirements and harvest expectations. Agricultural systems that grow many crop types together are much more complicated to measure, but measuring productivity of these operation is important because this is the primary form of agriculture practiced by small-scale operations and in urban settings. This study was initiated to contribute to the body of knowledge related to urban agricultural productivity.

People who grow food, whether in cities or in rural areas, are usually concerned with *how much* they can grow. There are many ways that this question can be answered, depending upon the perspective it originates from. For example, home food gardeners may be most concerned with growing types of fruits and vegetables that are important to their food heritage. In this scenario, growing enough culturally important produce is what is important. A subsistence food producer is concerned with growing a variety of crop plants that provide for a healthy diet. A commercial urban farmer may be concerned with growing enough so that the business is economically sustainable. From a broader perspective, the amount that can be produced within a city's greenspace may be of interest to community leaders and policy makers who wish to reduce dependency on food grown elsewhere. Even though there are a variety of perspectives, the one common element is that *the food growing operation must yield enough benefit to be worthwhile or the effort will eventually be abandoned.*

Since many urban food-producing operations are non-commercial, answering the question of "how much can be produced?" is more complicated than for commercial operations that have clearly-defined products, valuations of those products and accounting of profits. Home food gardens and community gardens are not usually preoccupied with growing crops that have high market value or a large profit margin. Instead, these food gardens are planted to provide fruit and vegetable varieties that may be culturally important to the grower or for other personal reasons. Because of this, the definition of a "crop" is much broader than for commercial crops.

This residential food growing operation study will address the question of "how much can be produced from a small-scale residential food growing operation," but only *one* growing operation. Because this study represents a single data point, it cannot be used to calculate the kinds of statistics that require many data points collected across a range of growing conditions. However, this case study can guide others who wish to collect and analyze similar data so that the importance of urban agriculture can be more broadly quantified and understood.

What Are the Costs and Benefits?

Was it worthwhile to grow food? The term "worthwhile" can carry many meanings, depending on who is asking the question, what perspective they come from and what they would define as being worthwhile. The sustainability scientist approaches this question from the perspective of the three pillars or sectors of sustainability: environmental, economic and social. These sectors are often viewed as separate entities but they are very much blended (Figure 1); more about how they were measured in the case study is provided below.

Environmental and Ecological Costs and Benefits

For a food growing operation to be environmentally sustainable, it must not damage or deplete the resources upon which it depends. By extension, the operation may not negatively affect the land it occupies, adjacent lands or downstream areas. Recall that sustainability, for some, is viewed as being a binary (or absolutely defined) value- something either meets a criteria ("...must not damage or deplete the resources upon which it depends") or it does not. For others, a specific activity is considered to be more sustainable than another, by comparison, because it uses limited resources more efficiently and causes less environmental impact. The residential food growing operation study was not able to measure absolute sustainability because not all aspects of environmental, economic and social sustainability could be directly measured. However, it is able to measure how much more sustainable the operation was in comparison to other alternatives.

The environmental parameters that were quantified in this study included water (use and conservation), energy (use and conservation), CO_2 emissions (produced and reduced) and waste materials (generated and recycled). These were calculated using the best available information at the time of the printing of this book.

Urban landscapes are highly modified environments that have had most of the original ecological community stripped away during the development process. Making cities more environmentally functional requires improving the urban ecology. Some of the questions this study sought to answer were:

- What were the ecological aspects of the food garden?
- How did the food production site function as wildlife habitat?

Economic Costs and Benefits

For a food growing operation to be economically sustainable, it must be profitable while not damaging or depleting the resources upon which it depends. This study measured the financial benefits of the food operation to determine if it was economically worthwhile. This question is important to many urban food growers, especially those that face space limitations or who are unable to produce large quantities of food. Although profitability may imply financial gain, this is not always the case. A successful community garden or home food garden may be defined by the amount of produce that is grown or the intangible benefits the garden provides.

Social and Intangible Costs and Benefits

What social benefits resulted from the residential food growing operation? Social sustainability is often described in terms that include intangible elements and quantitative metrics exist to measures some of those elements. Many practitioners of urban agriculture have identified a variety of intangible benefits that were important reasons for growing food. Although these intangible benefits often difficult to quantify, a survey was conducted to understand local residents' attitudes and perspectives about urban food growing and how this operation may have benefitted the community.

Can Enough be Grown to Meet Residents' Needs?

Can enough food be grown within cities to make a significant contribution to urban residents' food needs and food sustainability? At the household level, the answer to that question may be dependent on how much is space available for growing and the productivity potential from that space. From a community-level perspective, one must know how much can be grown within the urban footprint to reduce or eliminate the local food demand. Results from this study were extrapolated to the community-level to examine this question, while respecting the limited application of this relatively small data set.

Urban Agriculture and Food Sustainability

The concept of food sustainability was introduced in the previous chapter. Calculations of the local food demand, potential food production and food sustainability index were conducted from study data. From these analyses, a clearer understanding of the role that urban agriculture can play in meeting local food needs was obtained.

Study Description

This study measured a broad range of costs and benefits that are associated with small-scale urban food production. A list of these factors is shown in Table 18, which includes those that can be considered to be benefits or costs, depending on how they are consumed or produced during the food growing operation. The focus of data collection efforts in the residential food garden study was adapted over the study period to accommodate new information and insight that was gained through time. Over the case study's five-year period, sub-studies were conducted to further investigate certain aspects that were found to be important. In some cases, these sub-studies were initiated because of a fortunate set of circumstances that permitted data to be collected when it would otherwise not have been possible. These sub-studies could not be conducted over a long period of time, but nonetheless were informative with respect to the magnitude of benefit or cost of a particular factor.

Study Site

The food growing operation study was conducted in a residential neighborhood in the West Palm Beach metropolitan area of south Florida, U.S.A (Figure 10). The single-family home's residents consisted of two adults, both with full-time careers and limited time to dedicate to food growing. The food growing operation was initiated and maintained by one member of the household and the second member of the household did not participate in the food growing activity. The person responsible for the food garden had a low-level of skill with respect to growing food

Table 18. Factors measured in the food growing study, including examples of how each may be a cost or benefit.

Factor	Classification	Description
Water	Cost	Consumed during the food growing operation
	Benefit	Savings as compared to other urban land cover types
Energy	Cost	Consumed during the food growing operation
	Benefit	Savings as compared to purchased food
Greenhouse Gas Emissions	Cost	Emissions resulting from the food growing operation
	Benefit	Fewer emissions as compared to other uses for the growing space
Waste	Cost	Generated during the food growing operation
	Benefit	Assimilated during the food growing operation
Fertilizer	Cost	Consumed during the food growing operation
	Benefit	Capture of compostable material to recycling of nutrients
Productivity	Cost	n/a
	Benefit	Production of food
Capital/Finances	Cost	Investment in the operation, both startup and maintenance
	Benefit	Reduction of essential food purchases
Labor	Cost	Investment of labor to start and maintain operation
	Benefit	Physical benefits from exercise
Ecology	Cost	Impacts from resources consumed by food growing operation
	Benefit	Importance of food growing site to the urban ecology
Social	Cost	Household and community-level impacts
	Benefit	Household and community-level benefits
Sustainability	Cost	n/a
	Benefit	Increases urban food sustainability

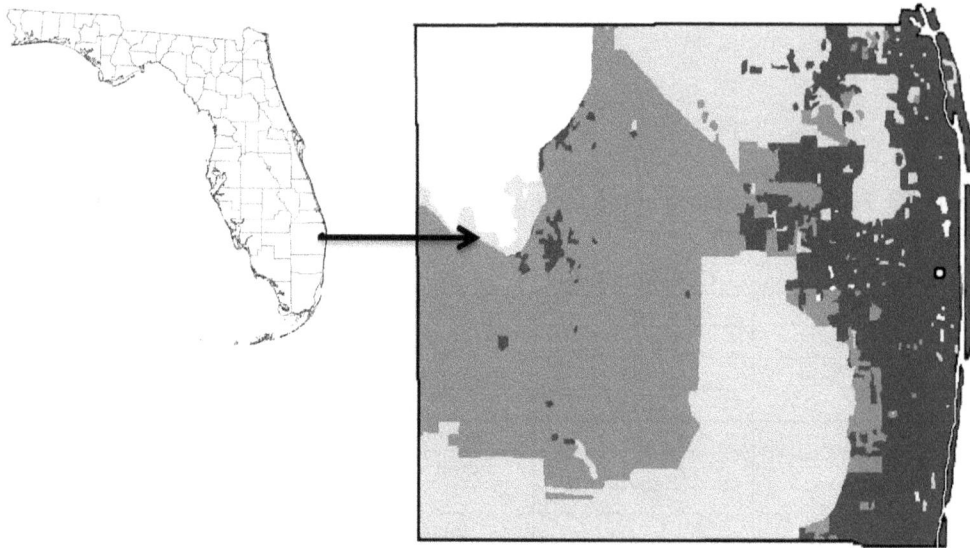

Figure 10. Map of the Palm Beach County metropolitan area in south Florida, USA. Study site location indicated by black circle. Major land use types (c. 2009) are indicated as: white=water, light gray=undeveloped land, medium gray=agricultural land and dark gray=urban land (Based on data source: SFWMD 2009).

crop plants in the study area. Because of this, the first two years of the study required a substantial investment in learning which crop plants were suitable to grow in the region and how to care for them.

The neighborhood was founded in 1957 and consisted primarily of single-family residences. The residential property where the study was conducted was approximately 0.1 hectare (0.25 acre) in extent and contained one single family home of approximately 130 m^2 (1,400 ft^2) a utility shed of approximately 9 m^2 (100 ft^2), a small concrete patio, sidewalks and a concrete driveway (Figure 11). Soil at the study site was a level Arent (United States Department of Agriculture Soil Conservation Service 1978), well-drained mostly sterile sand of marine origin.

The West Palm Beach metropolitan area occupies the eastern portion of Palm Beach County. There are several communities in the county that are outside of the metropolitan area, but the total population of those communities is small relative to the metropolitan area. Palm Beach County, and particularly that of the metropolitan area, experienced large increases in population between the 1970s and 2014 (Table 19). When constructed, the neighborhood was on the urban fringe but is now considered to be a part of the urban core.

Figure 11. Location of the food growing study site within an urban residential neighborhood (indicated by black square). Image source: 2014 Palm Beach County Digital Ortho Quad, 15 cm (0.5 ft.) resolution.

Table 19. Palm Beach County population from 1970 through 2014.

Year	Population*
1970	348,753
1980	576,863
1990	863,518
2000	1,131,184
2010	1,320,134
2014	1,397,710**

*Source: Decennial Census, U.S. Census Bureau
(http://www.census.gov/prod/www/decennial.html)
**http://www.census.gov/popest/data/cities/totals/2014/SUB-EST2014.html

At the time of the study, Palm Beach County was Florida's largest agricultural producer and led the nation in the production of sugarcane, fresh sweet corn, sweet bell peppers and was also the state's largest producer of rice, lettuce, radishes, Chinese vegetables (e.g., bok choi, Chinese cabbage), specialty leaf vegetables and celery (U.S. Department of Agriculture, 2009a). The extent of agriculture in Palm Beach County has declined steadily since the 1980s (Figure 12), concurrent with a rapid rise in the county's population. There were 301,000 hectares of farmland in 1978 (U.S. Department of Agriculture 1981), which fell to only 186,200 hectares by

165

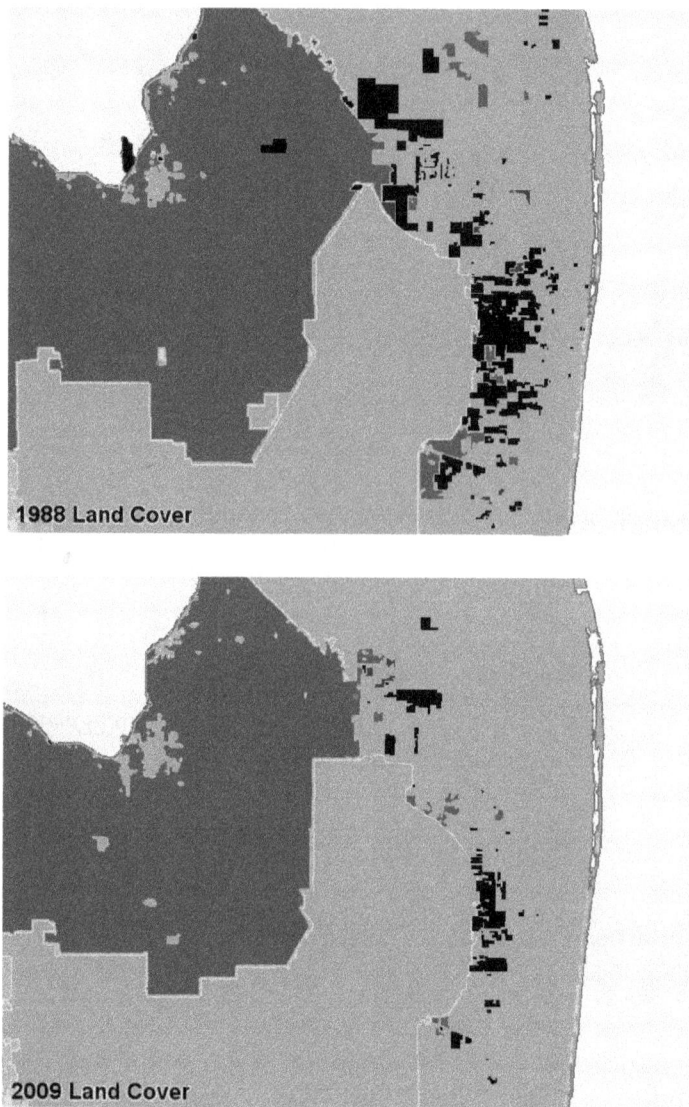

Figure 12. Extent of agricultural lands in Palm Beach County, Florida in 1988 and 2009; white=water, medium gray=urban and undeveloped land, dark gray=sugar cane, black=groves and truck crops (e.g., vegetables, melons). Data source: SFWMD, 1990; SFWMD 2009.

2009 (SFWMD 2009). The land dedicated to vegetable production fell from 48,560 hectares in 1978 (U.S. Department of Agriculture 1981) to 32,290 hectares in 2009 (SFWMD 2009).

The West Palm Beach metropolitan area is the northern portion of the sprawling Miami-Fort Lauderdale-West Palm Beach FL Metropolitan Statistical Area as defined by the United States Census Bureau. This metropolitan area encompasses the developed areas of Miami-Dade, Broward and Palm Beach counties. The estimated population of the metropolitan area in 2010 was 5.6 million permanent residents. The metropolitan area is constrained by the Atlantic Ocean to the east and the remnant Everglades to the west (Figure 10).

West Palm Beach has a tropical rainforest climate according to the Köppen climate classification system. The region has two distinct seasons: a cool, dry season that occurs from November through April and a hot, wet season that occurs from May through October. The onset of these seasons can vary each year, making October and May transitional months with unpredictable rainfall conditions (Table 20). The region also has a hurricane season that lasts from June 1 through November 30.

Study Duration

The study was conducted from 2009 through 2015. (Table 21). The length of the study was not fixed at the beginning; instead, a decision to continue with an additional year of data collection was made at the end of each growing season. Factors that played into this decision included: (1) the environmental conditions over the previous growing season, (2) adjustments to the study's focus, and (3) the quality of data collected over the previous growing season. The first year of the study included a growing period that was shorter than the following four years (Table 12).

South Florida's year-round growing season can be divided into two parts based upon the crops that are being grown: one is the temperate crop season when "winter" vegetables are grown (see below) and is comparable to the growing season in some temperate climate regions. The second is the tropical crop season, which spans most of the year. The normal period for growing fruits and vegetables associated with mid-latitude climates is from October 15 through March 15. Although one can plant temperate-zone vegetables outside of this period, they require a great deal of care and control of environmental conditions to produce. By April 1st, the sun's strength and prolonged dry conditions make it difficult to keep temperate-zone crop plants productive. Additionally, crops grown during the summer season may be damaged or destroyed by tropical cyclones (tropical storms or hurricanes).

Table 20. Climate data for West Palm Beach, Florida (1981-2010, source: NOAA data from Palm Beach International Airport).

	Mean high °C (°F)	Record high °C (°F)	Mean low °C (°F)	Record low °C (°F)	Mean rainfall mm (inches)	Mean rain days (> 0.01 mm, in)
Jan	24 (75)	32 (89)	14 (57)	-3 (26)	80 (3.1)	8.1
Feb	25 (77)	32 (90)	15 (59)	-3 (27)	72 (2.8)	7.6
Mar	26 (79)	35 (95)	17 (62)	13 (26)	117 (4.6)	8.9
Apr	28 (82)	37 (99)	18 (66)	3 (38)	93 (3.7)	7.1
May	30 (86)	37 (99)	22 (71)	7 (45)	115 (4.5)	9.7
Jun	31 (88)	38 (100)	24 (74)	16 (60)	211 (8.3)	15.3
Jul	32 (90)	38 (101)	24 (76)	18 (64)	146 (5.8)	15.1
Aug	32 (90)	37 (99)	24 (76)	18 (65)	202 (8.0)	17.4
Sep	31 (88)	36 (97)	24 (75)	16 (61)	212 (8.4)	16.7
Oct	30 (85)	35 (95)	22 (72)	8 (46)	130 (5.1)	12.1
Nov	7 (80)	33 (92)	19 (66)	2 (36)	121 (4.8)	9.2
Dec	25 (76)	32 (90)	16 (60)	-4 (24)	86 (3.4)	8.7
Year	28 (83)	38 (101)	20 (68)	-4 (24)	1583 (62.3)	136

Table 21. Data collection periods in the study.

Season	Produce Grown
October 1, 2009 – June 6, 2010	Vegetables
September 1, 2010 – August 31, 2011	Vegetables & fruit
September 1, 2011 – August 31, 2012	Vegetables & fruit
September 1, 2012 – August 31, 2013	Vegetables & fruit
September 1, 2013 – August 31, 2014	Vegetables & fruit
Summer 2015	Fruit

Other challenges with growing temperate vegetables in south Florida are the light and temperature regimes. In the mid-latitudes, the length of day is longest during the growing season; however in south Florida the growing season occurs during the shortest days of the year. Temperature regimes are quite different too. In temperate climates, the season begins cools in the spring, then warms, and then cools in the fall. In southern Florida, the season begins hot, then turns mild to cool, then becomes hot again. These differences in light and temperature regimes provide less-than-optimal growing conditions for vegetables that originate from the temperate latitudes.

The annual growing season (including both the temperate crops and the tropical crop parts), as defined in this study, began September 1st and continued for a full year, concluding on August 31st. Study period began at the onset of the winter vegetable growing season in south Florida, which is the main food growing season for this region

Food Growing Site

The food growing site consisted of a residential backyard garden plot, an adjacent "wildlife garden," utility areas (composting, storage, etc.) and paths for access (Figure 13). Besides this space, there was a vertical garden on the back wall of the house and a large mango tree in the front yard of the house. The extent of food growing space during the study varied from year-to-year because site conditions changed through time. For example, tree and hedge growth changed light and soil moisture regimes in some production areas and, through time, rendered these unsuitable for crop growth. The total backyard study footprint area and growing space utilized each year are shown in Table 22. To compare productivity between years, values will be reported in terms of the amount of food produced per area of growing space.

The wildlife garden consisted of a variety of plantings of native and ornamental species that were used as an alternative to lawn, but did not produce food. This garden provided a permanent habitat for agriculturally beneficial species that would benefit the cultivated food plants. A large, mature live oak (*Quercus virginiana*) tree was part of the wildlife garden and provided shade to the vegetable garden late in the afternoon.

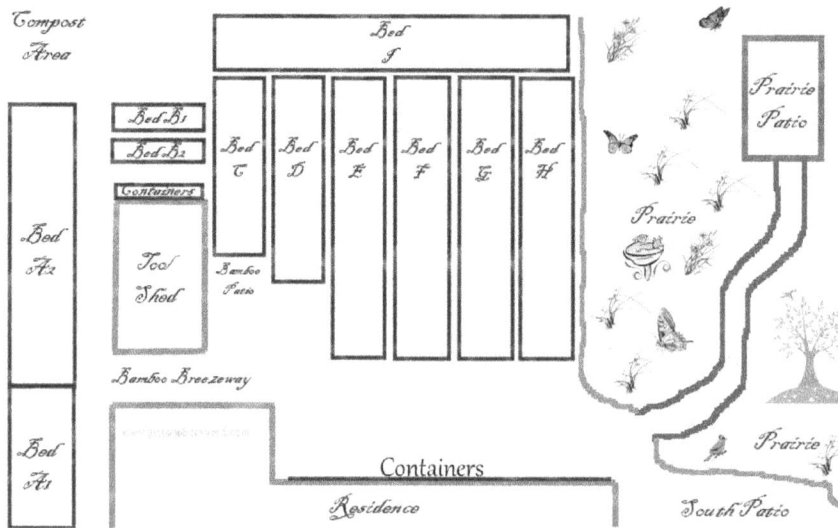

Figure 13. Map of the residential food growing study site showing production bed locations and the adjacent wildlife garden. Vertical wall garden is indicated by "Containers".

Data Collection Methods

Environmental Parameters

Environmental data, including rainfall, temperature and extreme weather events) were obtained from the National Weather Service's weather monitoring station at Palm Beach International Airport, which was located 5.1 km (3.2 miles) from the study site. Weather data statistics were recorded for each study year and included mean, minimum and maximum temperatures, duration of dry periods, as well as the duration of extreme hot and cold events.

Water

During the study period, irrigation water was supplied from three sources: the home well, a municipal water supply and captured rainwater. The volume of well water applied to the garden was calculated by multiplying the amount of time the pump ran with the pump output (liters per minute, or lpm), which was measured monthly to verify output. Irrigation water from the municipal water source was calculated by multiplying the amount of time spent watering by the spigot output (lpm). Spigot output was measured monthly to verify output.

Table 22. Space partitioning during the backyard food production study; all values are m^2.

Season	Growing Space	Utility	Other	Total Footprint
2009-2010	[1]Vegetables: 70 [2]Fruit: 7	[3]Compost: 2 [4]Mulch: 2 [5]Materials: 2	[6]Paths: 18	101
2010-2011	Vegetables: 100 Fruit: 7	Compost: 2 Mulch: 2 Materials: 2	Paths: 25	138
2011-2012	Vegetables: 65 Fruit: 7	Compost: 2 Mulch: 2 Materials: 2	Paths: 20	98
2012-2013	Vegetables: 65 Fruit: 7	Compost: 2 Mulch: 2 Materials: 2	Paths: 20	98
2013-2014	Vegetables: 74 Fruit: 7	Compost: 2 Mulch: 2 Materials: 2	Paths: 21	108
2015	Fruit: 7	n/a	n/a	7
2011-2014 Mean	Vegetables: 68 Fruit: 7	Compost: 2 Mulch: 2 Materials: 2	Paths 20	101

[1]Vegetables: the total area of garden bed space used to grow vegetables including in-ground, container and vertical gardening spaces; vertical growing space was approximately 2.8 m^2 from 2011-2014.
[2]Fruit: the total area of space dedicated to productive fruit trees.
[3]Compost: the area dedicated to composting, including a compost bin and compost storage.
[4]Mulch: the area dedicated to storage of mulch.
[5]Materials: the area dedicated to storage of garden supplies.
[6]Paths: the area of pathways between and around garden beds.

Over the study period, rain barrels were installed to store captured rainfall runoff from the house's roof. Initially, two 208-liter (55-gallon) capacity food-grade plastic barrels were installed during the second year of the study. During the third year, two additional barrels were installed to increase storage capacity. Captured rainwater that was used for irrigation was measured as it was applied using a hand-watering bucket with a fixed capacity of 9.5 liter (2.5 gallons).

Energy

Electrical energy consumed during the study was primarily associated with the operation of a 1.5 horsepower well pump used to extract groundwater for irrigation. The amount of energy consumed by the pump was estimated to be 1811 watts (based on manufacturer's amperes and voltage ratings with a power factor of 0.75), which was used to calculate the total electrical kWh used each year of the study. Total annual electricity use (kWh) associated with the irrigation pump's operation was calculated from the time the pump was in operation (recorded during the study period) and the energy consumed by the pump (1811 watts). Climate-change related damage associated with electricity usage was calculated Natural Resource Council's (2010) values reported in Table 15, with electrical power from the utility company assumed to be generated from natural gas.

Energy costs associated with transportation, which were primarily related to the consumption of fossil fuels and the subsequent release of greenhouse gasses, arose from the acquisition of materials used to construct and maintain the food-growing site. The distance that was traveled by a light-duty vehicle to acquire building materials, fertilizers, compost, mulch, horse manure, organic pest control products and other supplies was recorded over the study period. The greenhouse gas and non-greenhouse gas impacts, as well as the amount of CO_2 emissions that were released from burning fossil fuels, was estimated from the distance traveled using the light-duty vehicle (see Chapter 9).

Because the food produced for the household was being grown locally and there were no shipping-related energy costs, the food garden produced a net benefit by reducing fuel consumption (as compared to not growing the food and purchasing it from a retail market). The result from this analysis yielded an estimate of the amount of crude oil saved by growing the food locally rather than purchasing the same food from a retail store. This benefit was calculated by:

1) Weighing the total amount of each type of fruit and vegetable harvested from the garden each month

2) Surveying grocery stores in the area monthly to determine where these same fruits and vegetables would have come from if they had been purchased from a retail outlet (see below)

3) Calculating the amount of energy that would have been expended to truck the harvested produce the (weighted) average distance that similar food items in local grocery stores had been shipped; the energy units are gallons of diesel fuel; semi-truck fuel consumption data (see below).

4) Calculating the amount of petroleum that was not expended to transport the harvested produce the average distance that similar good items in the local grocery store had been shipped (see below)

Calculating the average distance that vegetables and fruit in the grocery store had traveled was straightforward. Each month, the origin of harvested vegetable and fruits types offered for sale in grocery stores were recorded and a weighted average for all product was calculated. The shipping distance was used to estimate the greenhouse gas and non-greenhouse gas impacts (see Chapter 9), as well as the amount of CO_2 emissions that were released from burning fossil fuels during transport.

An alternative method that could be used to estimate transportation costs related to food is to rely upon published sources that indicate that food in grocery stores has traveled is approximately 2,400 – 4,025 km (1,500 - 2,500 miles) (Florida Fish and Game Commission & The American Farmland Trust 1995, Worldwatch Institute 2016). For this study, the data recorded from grocery stores was used rather than these estimates.

The amount of fuel needed to transport produce the average distance that purchased food would have traveled was calculated. This fuel volume was estimated by dividing the average distance the food would have traveled (if purchased and not grown locally) by the average liters/km (miles/gallon) fuel consumption rate of a typical semi truck. The American Transport Research Institute (2009) and the Environmental Defense Fund (2017) estimated the average fuel efficiency for a semi tractor-trailer to be 8-10 km/gal (5-6 miles/gal.), excluding the additional weight of a payload. These values were used in this study.

The amount of diesel fuel that would have been consumed if the grown food had been transported (i.e., the average distance that comparable produce in the grocery store had traveled) was converted into number of barrels of crude oil. This conversion was based on information from the U.S. Energy Information Administration, which indicated 1 barrel of crude oil would yield about 19 gallons of motor gasoline and 12 gallons of diesel fuel (U.S. Energy Information Administration 2016).

In the same way that there are climate-change related damages associated with electricity consumption, there are also climate damages associated with long-distance food transportation that consumed fossil fuels. Since the residential food garden avoided the consumption of fossil fuels, it also avoided the associated climate-change damages and this can be calculated using estimates from the Natural Resource Council (2010). This was calculated using the average number of kilometers that the food would have traveled had it been purchased and not grown locally. The average non-climate damages per kilometer used in this study can be found in Chapter 9.

Waste

The amount of waste material generated and assimilated during the study was recorded by type in the following way:

- Mulch was obtained from a local waste management organization that collected residential landscape waste material (tree, shrub and lawn clippings) and turned it into mulch for public use. The volume of mulch that was obtained and applied to the garden was recorded.

- Compost was obtained from a local waste management organization that collected residential landscape waste material and turned it into mulch for public use. The volume of mulch that was obtained and applied to the garden was recorded.

- The amount of landscape waste material generated by the household and applied to the garden as mulch was relatively small and was not recorded.

- Over a period of 276 days (2011-2012 season), compostable organic waste material was picked up daily (weekdays only) from a commercial cafeteria kitchen. The material was predominantly fruit and vegetable scraps. Each day that compostable material was acquired, the volume and weight were measured and recorded before adding to the compost bin. After the material was decomposed into compost, this was added to the garden as a soil amendment and organic fertilizer.

- The amount of compostable kitchen waste generated by the household and applied to the garden was measured (weight and volume) and recorded over the study period.

- The amount of waste generated by the household and set out for landfill pickup (non-recyclable materials) was measured (volume and weight) over a period of 7 weeks during the 2011-2012 season.

During the study period, the volume of mulch and compostable materials that had been assimilated by the food growing operation was measured. Normally, this waste would have gone to a landfill but was recovered and used as an essential input into the food garden. The amount of money that had been indirectly saved by not disposing this waste material was estimated from fees charged by the local waste management organization. This fee, as of 2016, was $42.00 per ton.

Besides the landfill disposal of waste food scraps, there are also greenhouse gas emissions that result from burying the material in a landfill. The amount of methane produced by this disposed material was estimated from the US Environmental Protection Agency's WARM model's table of estimated methane production from disposed materials (US Environmental Protection Agency 2014). The production rate for food scraps was 399.5 m^3 per dry mg. Because the weighed food scraps recovered during this study were wet weight, it was assumed that the dry weight of this material would be much less. A review of Internet resources indicated that certain types of fruits and vegetables lose approximately 50 to 80 percent of wet weight during dehydration. Based on these sources, this study will assumed that the dry weight of the recovered compostable material would be 35 percent of its wet weight.

Fertilizers

The residential food garden study relied exclusively on purchased organic fertilizers and compost as the primary sources of plant nutrients. These each came with costs- the purchase fertilizer required a cash expenditure and the compost required time investments to acquire or create. The total cost of purchased fertilizers was recorded during each year of the study. The total amount of time dedicated to acquiring compost or creating compost was recorded as part of the labor costs of the operation.

Because the purchased organic fertilizers and compost used during the study were slow-release fertilizers that carry a low risk for pollution runoff, there is a net benefit to the environment by avoiding these impacts. The Natural Resource Council (2010) estimated that pollution from farms in the United States cause an estimated $120 per hectare ($298 per acre) (2016 USD) in non-climate damages each year. By growing food locally using slow-release types of organic fertilizers, this environmental cost was avoided and is considered to be a net benefit caused by the residential food growing operation. The total benefit realized each year from this operation was calculated by multiplying the area in cultivation by $0.012, the non-climate damages expressed in 2016 USD per m^2 of farm area.

The production of some chemical fertilizers has caused environmental degradation or destruction. For example, phosphate mining for fertilizer has destroyed hundreds of thousands of hectares of natural habitat and the use of chemical phosphate fertilizers carries these environmental costs. Although the currency value of these impacts have not been quantified, it is important to consider these impacts along with the use of chemical phosphate.

Productivity

The measured weights of all harvested fruits and vegetables (by variety) were recorded daily. All weight measurements were taken on a Chefmate 3 in 1 digital scale whose calibration was verified monthly using a set of standard weights (1, 2, 5, 10 and 20 gram chrome calibration weights). These harvest data were converted to number of cups (servings) of five types of vegetables and fruit described by the Choose My Plate guidelines (Table 12). *It is recognized that the other food groups (grains, protein foods and dairy) are important for a balanced and healthy diet, however their production was not a part of this study and will not be further considered.*

The conversion between the harvest weights for a single crop type into cups was accomplished by sampling harvested produce to determine how much (by weight) of a type of vegetable equated to a 1-cup volume. Several samples were taken over the duration of the study, when possible. For some types of vegetables, a single conversion factor was used for several types of vegetables that had similar physiological characteristics (e.g., spinach and leaf lettuces). Besides these measured conversion factors, other published sources (e.g., Choose My Plate guideline descriptions) were consulted. It is recognized that this conversion from weight to serving cups could be a potential source of error because of variations in the density of produce from season to season. These estimates were generated for the purpose of calculating a gross measure of how closely production could match dietary demands and were not intended to be precise values.

Harvest Valuation

The total amount of produce harvested, expressed as weight and crop type, was continually recorded (at harvest) over the study. The values of all harvested crops were determined by conducting a monthly survey of retail prices at local grocery stores. All of these stores were franchise supermarkets that were among the largest food retailers in the metropolitan area. These included:

- Publix Supermarket (specialized in conventionally-grown produce)
- Winn Dixie Supermarket (specialized in conventionally-grown produce)
- Walmart Supercenter (specialized in conventionally-grown produce)
- Publix Greenwise Market (specialized in organically-grown produce)
- Whole Foods Market (specialized in organically-grown produce)

When specialty produce items were not available at these large food retailers, smaller specialty markets were surveyed. At the onset of the study, prices were surveyed several times per month; but it was later determined that produce prices did not vary significantly from week to week. Because of this, produce prices were surveyed monthly. Two produce values were recorded for each crop type-conventionally grown and organically grown produce. Because the garden study used organic methods, the actual valuation of harvest is in line with the organic produce prices.

It was not always possible to obtain prices for both conventionally grown and organically grown produce for all crops varieties that were harvested during the study. In some cases, prices for conventionally grown produce were available, but prices for organically grown produce were not. In other cases, prices for a specific type of produce were not available; this was particularly true for specialty produce. Methods were developed to reasonably estimate prices that were not available by the supermarket produce department surveys. In cases where prices for a specific type of produce were not available, a surrogate price from a comparable type of produce was used in its place. A surrogate had the following characteristics in common with the unavailable produce type: (1) similarities in the part of the plant used for food, such as leaf or fruit types, (2) they are both typically prepared in the same manner, and (3) they are sold using the same units, such as weight or volume.

In cases where either conventionally or organically grown produce prices were available (but not both), the missing price value was estimated using a conversion factor between the two. The conversion factor was calculated from price data collected during the study period and the mean percent difference between the conventional and organic prices obtained from the stores.

Labor

All time spent working on the residential food garden was recorded for each of the following categories:

- Gardening, which in included soil preparation, planting, weeding
- Time spent acquiring mulch and compost from outside sources
- Time spent working on maintaining a composting operation
- Watering
- Construction and maintenance of garden infrastructure
- Pest and disease control

Expenditures

The cost of all materials (cash expenditures) was recorded over the study period and summed by year. Costs included purchases of building materials, seeds, organic fertilizers, potting soil, soil amendments, organic pest control products, plant supports, electricity to run the well pump, etc. These costs were groups by similar types and the total for each group were summarized.

Ecological Parameters

The food-growing site occupied the backyard of a residential property. When the homeowner originally purchased the property, a decision was made to break with the traditional irrigated lawn landscaping, which was typical for the neighborhood, and to plant low-maintenance landscape and a large food-producing garden instead. For the most part, this property was an exception to the neighborhood lawnscape and the ecological data collected from this site can be helpful to understand how replacing irrigated lawn with food gardens can be beneficial to the urban environment.

Throughout the study period, wildlife observed within the garden and the adjacent wildlife garden was recorded. Wildlife included birds, land animals and invertebrates (e.g., insects). Species lists were kept, but abundance was not measured.

A plant inventory was recorded over the study period to document the uncultivated species (weeds) growing at the site. These plants were researched in the scientific literature to determine if they were used as food plants in other cultures or if these plants were known to be larval or nectar plants for organisms such as bees, butterflies, birds and other agriculturally beneficial organisms. The inventory contained a species list, but abundance was not measured.

Urban development contains a large amount of greenspace, some of which may be suitable for food production- even in planned residential communities (Figure 14). An analysis was conducted to examine the area potentially available to grow food within residential neighborhoods of metropolitan Palm Beach County. Only areas within residential property that were located in side yards and backyards, away from principle street view and not obstructed by trees or structures, were included. The reason that front yards were not included is that many communities in the area have bylaws or community ordinances that require maintained lawns in the front of the home; these laws are intended to preserve the neighborhood's appearance.

An analysis was conducted to visualize the extent of lawn in the neighborhood where the food-growing site was located using aerial photography and a digital image editing program (Paint Shop Pro v. 7). This analysis was valuable to understand how urban greenspace was distributed across the developed area. The image was processed using high saturation and contrast controls to delineate borders between features. Image pixels that were associated with trees, as well as buildings, roads swimming pools and other development features, were manually removed (visualized as black). The remaining pixels, which represented the extent of low-profile vegetation, were displayed using a dark to light green color gradient. Field verification of the digitally edited image was conducted and corrections were made to the image to ensure accuracy. The final verified processed image represented the extent of lawn in the neighborhood.

Social Parameters

The importance of the food garden to the household, as well as other social factors, were identified during the study but these were not quantitatively measured. Beyond the household, the importance of residential food gardens to local residents was the subject of a community survey that was conducted from January through March 2013. That study sought to:

- Identify the ways that urban food gardens are important to local residents
- Describe respondents' attitudes towards and feelings about urban food gardens
- Understand why local residents grew food for themselves
- Identify the intangible benefits and costs reported by study participants

Figure 14. Residential urban development in metropolitan West Palm Beach, Palm Beach County, Florida, USA. [Image source: 2014 Palm Beach County Digital Ortho Quad, 15 cm (0.5 ft.) resolution].

The community survey also explored how common the practice of residential food growing was in the area, the amount that people grew and respondents' food growing heritage. More than 275 respondents filled out detailed questionnaires that asked about their experiences with growing food, the history of food growing in their family, demographic information and their attitude about growing food for themselves. Of these respondents, 45 participated in a personal interview that further explored their food growing background, their feelings about the practice and its importance to them. Details about the background and methods used in that community survey can be found in Zahina-Ramos (2013).

Potential Residential Food Production in the Metropolitan Area

Two questions of interest to urban agriculture and sustainability scientists are: (1) how much food is being or could be produced within a city and (2) how much does this food contribute to residents' food needs. The results from the residential food garden study provided data of sufficient granularity to allow a tentative examination of this question and, in doing so, it can represent an example of how these questions can be quantitatively measured by other investigators. By extrapolating the production results from the residential food garden study, the potential food production capacity of the metropolitan area was calculated. It is necessary to understand that the case study is a single study point and does not representative of all growing conditions or levels of productivity that could be realized by food growing operations in the area. However, this analysis is informative and can provide insight into how residential food gardens can contribute to local food production.

This analysis was conducted using the following steps, each of which is described in detail below:

1. The amount of residential food growing space that was available within the West Palm Beach metropolitan area was estimated using high-resolution aerial photography
2. The Potential Food Production (FP_P) from available residential growing space was estimated using production data from the case study
3. Census data from the metropolitan area were used to calculate the Potential Food Demand (FD_P)
4. The Food Sustainability Measure (FS_M) was calculated for the metropolitan area; because the extent of the metropolitan area was considered to be the foodshed (for the purposes of this analysis), the FS_M is useful to understand the potential foodshed efficiency of the metropolitan area
5. The Food Sustainability Index (FS_I) was calculated for the metropolitan area

Estimating Available Residential Food Growing Space

The potential area of residential food growing space within the Palm Beach County Metropolitan Area was estimated from land use maps and digital aerial photos that were processed in ArcGIS (v.10.3). The extent of the metropolitan area was derived from the 2008-2009 Land Cover Land Use (LCLU) shapefile obtained from the South Florida Water Management District's on-line GIS Data Catalogue (SFWMD 2009, available from www.sfwmd.gov). All urban land cover types (residential, industrial, commercial and utilities) were selected and color-coded to visualize the extent of built land (Figure 10).

Because the amount of time and labor involved with identifying and measuring all potential food growing space within the metropolitan area was formidable, a sampling method was used to estimate the amount growing space. In ArcGIS, a shapefile of Florida Township-Range-Section lines, obtained from the Florida Geographic Data Library (www.fdgl.org), was overlaid on the LCLU map. This provided a grid of mostly equal sized cells (Sections), each being approximately 1 square mile in extent. Each Section was assigned a unique identifying number. Using a random number generator in MS Excel, twenty-five Sections were randomly selected within the metropolitan area, of which 15 fell within residential areas (Figure 15). Each of the 15 Sections was further divided into quarters (N, S, E and W according their cardinal direction), yielding a sample size of approximately one-quarter square mile (quarter-Section) in extent (Figure 16). One of the four quarters within the Section was randomly selected for analysis. These 15 quarter-Sections were assumed to be a representative sample of the residential development within the metropolitan area.

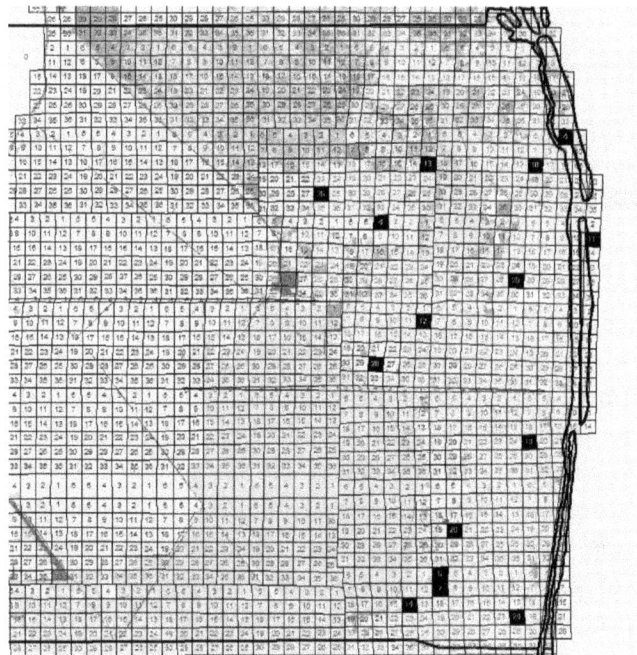

Figure 15. Township-Range-Section blocks in Palm Beach County indicating randomly-selected residential blocks (black) used in this study; black lines indicate county borders.

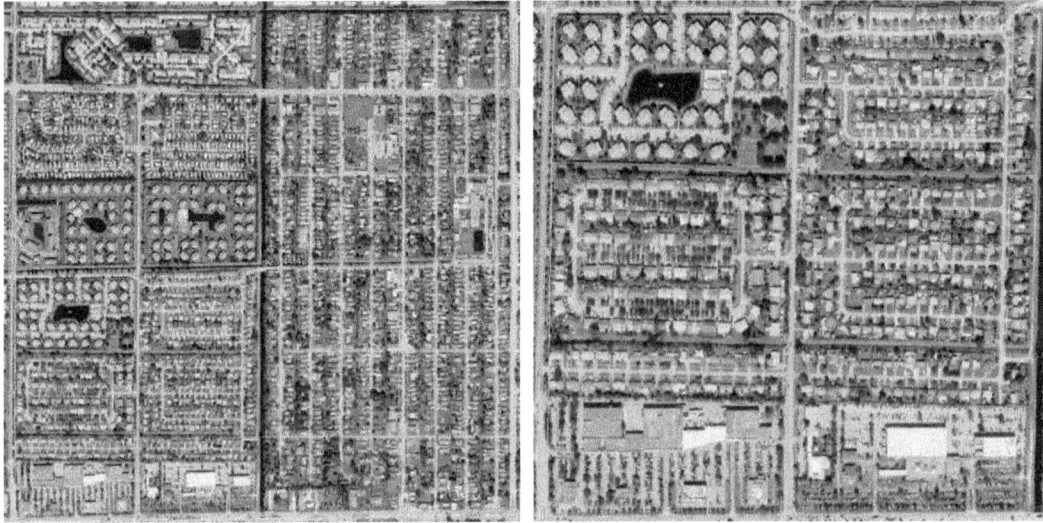

Figure 16. Sample residential Section (left) and SW quarter-Section (right) tract in metropolitan Palm Beach County, Florida, USA; [Image source: 2014 Palm Beach County Digital Ortho Quad, 15 cm (0.5 ft.) resolution].

Within each of the 15 quarter-Sections, the extent of residential greenspace that would be suitable for food growing were identified and measured. Aerial photography used in this analysis was obtained from Palm Beach County's digital aerial photography archives [2014 Palm Beach County Digital Ortho Quad, 15 cm (0.5 ft.) resolution]. Greenspace that would be potentially available for food gardens were visually identified on each residential lot and a spatial drawing tool was used to draw polygons to represented these areas (Figure 17). As previously mentioned only areas within residential property that were located in side yards and backyards, away from the view of principal streets and not obstructed by trees or structures, were considered as potential food growing sites because many communities in the area have bylaws or community ordinances that require maintained lawns in front of the home. The total potential food growing area within each the 15 quarter-Sections were summarized according to housing type (single-family, multi-family, etc.). The percent of potential food growing space within the quarter-Sections were used to estimate the total amount of potential food growing space in all residential land cover within the metropolitan area.

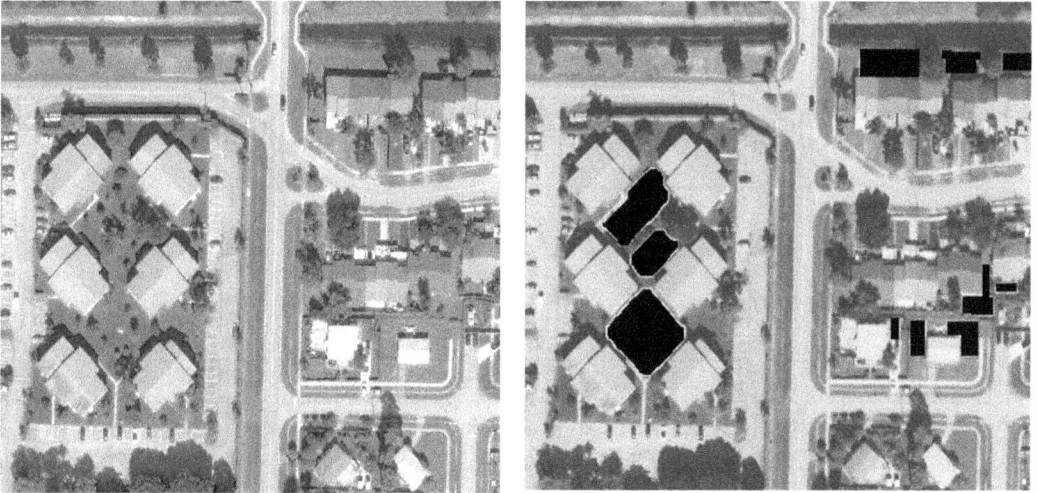

Figure 17. Left: sample aerial photo of a mixed multi-family and single-family residential community in metropolitan Palm Beach County, Florida, USA. Right: Potential food growing areas within the sample aerial indicated by black polygons [Image source: 2014 Palm Beach County Digital Ortho Quad, 15 cm (0.5 ft.) resolution].

Estimating Food Production Potential (FP$_P$) in the Metropolitan Area

The amount of food produced by the residential food garden was expressed as productivity per square meter for each growing season. This productivity was used to calculate the FP$_p$ possible from the available food growing area. The FP$_p$ was expressed as total weight and as cups (servings) for fruit and 5 types of vegetables (see Productivity section above).

Estimating Potential Food Demand (FD$_P$) of the Metropolitan Area

Food demand of the West Palm Beach metropolitan area was calculated by multiplying the number of residents in each demographic category by the recommended amount of food consumption per person (Table 17). The number of residents in the metropolitan area, as well as their respective genders and age groups, were obtained from the 2016 U.S. Census Bureau data (Table 23).

Table 23. U.S. Census demographic data for Palm Beach County (source: https://www.census.gov/quickfacts/table/PST045215/12099,1276600).

Demographic Categories		Percent of Total Population	Number of Residents	
			Female	Male
Female		51.7	746,450	697,360
Age	<5	5.1	38,069	35,565
	5-18	14.3	106,742	99,723
	18-65	57.6	429,955	401,679
	>65	23	171,683	160,393
Average household size: 2.54				

Chapter 13. Residential Food Growing Operation Case Study- Results

Climatic Conditions

Temperature and rainfall conditions that occurred during the study period are shown in Table 24. Record high temperatures were common during the first two years of the study. April & July 2011 were record hot months (6.4 and 3.2 °F above average, respectively) and August 2011 was the 5th hottest year on record at Palm Beach International Airport. The year 2011 was the warmest on record for West Palm Beach with 119 days of temperatures at or above 32 °C (90 °F) (mean was 56 days) and a streak of 46 consecutive days of temperatures of 32 °C or higher (July 8-August 23). (Palm Beach Post 2011).

Besides these unusually warm temperatures, crop plant productivity was impacted by unusually cool weather, which occurred during the dry season of 2010-2011. The National Weather Service announced that December 2010 was the coldest December on record for south Florida (National Weather Service 2010). Frost occurred over several nights in January 2010, December 2010 and January 2011, which damaged or killed many crop plants in the study garden plot.

Rainfall from late 2010 to early 2011 was well below normal and the U.S. National Drought Mitigation Center classified the region as being under Extreme Drought conditions. Mandatory water restrictions were put into place, which restricted the use of groundwater and surface water sources for irrigation. The period from October 1, 2010 to March 31, 2011 was the driest on record for the region with rainfall being 38 cm (15 in.) below normal for the period (Palm Beach Post 2011).

A hurricane impacted the study site in late August 2011. Hurricane Irene passed east of the Florida coastline with sustained winds of 193 km/hr. (120 mph), producing heavy rainfall and local flooding. The effects from hurricane Irene can be seen in the unusually high rainfall totals for the 2011-2012 season, which did not occur during the main growing time of mid-October through mid-March.

Table 24. Environmental conditions during the case study.

	2009-2010		2010-2011		2011-2012		2012-2013		2013-2014	
	Oct 1 – Jun 6	Oct 15 - Mar 15*	Sep - Aug	Oct 15- Mar 15	Sep - Aug	Oct 15- Mar 15	Sep - Aug	Oct 15- Mar 15	Sep - Aug	Oct 15- Mar 15
Mean Daily Temp.* °C (°F)	21 (70)	-	23 (73)	20 (68)	24 (75)	21 (70)	23 (73)	21 (70)	24 (75)	22 (72)
Mean Daily Min. Temp. °C (°F)	4 (39)	-	6 (43)	6 (43)	9 (48)	9 (48)	10 (50)	10 (50)	11 (52)	11 (52)
Mean Daily Max. Temp. °C (°F)	29 (84)	-	29 (84)	27 (81)	28 (82)	25 (77)	29 (84)	27 (81)	31 (88)	27 (81)
Heat days (≥32 °C)	13	13	31	0	15	0	8	0	39	0
Cold days (≤4 °C)	9	9	7	7	1	1	0	0	1	1
Min. Temp. °C (°F)	0 (32)	0 (32)	-1 (30)	-1 (30)	4 (39)	4 (39)	6 (43)	6 (43)	4 (39)	4 (39)
Max. Temp. °C (°F)	35 (95)	35 (95)	35 (95)	31 (88)	34 (93)	30 (86)	33 (91)	30 (86)	38 (100)	31 (88)
Rainfall* meters (inches)	1.01 (40)	0.69 (27)	0.96 (38)	0.20 (8)	2.06 (81)	0.43 (17)	1.63 (64)	0.30 (12)	1.52 (60)	0.48 (19)
Mean number of consecutive rainless days	-	9	-	6	-	8	-	6	-	6
Mean number of consecutive rainless days	-	23	-	25	-	25	-	31	-	20

Note that October 15 - March 15 corresponds to the growing season for temperate-climate crops in this region. Temperature and rainfall data from the SFWMD DBHYDRO Database: station FHCHSX (DBKEY V2453, temperature measured at 15-minute intervals and DBKEY V2452, daily mean temperature); station WPB AIRP_R (DBKEY16610, daily total).

Food Garden Productivity

More than 60 types of vegetables and 10 types of fruit were grown and harvested over the study period (Table 25). The total weight of produce harvested and productivity, by season, are shown in Table 26. Because the first two years of data collection were affected by unusual weather conditions, the mean harvested weight was calculated for the last three years of the study. Over the study period, the amount of vegetables harvested per unit area was comparable in the final three years of the study. One may argue that these data indicate that the upper limit of productivity had been reached, but this conclusion should not be accepted without further analysis since both the crop types and the amount of each crop type that were grown varied each season, making year-to-year comparisons tenuous. Fruit productivity varied through the years and did not follow a consistent pattern. This variability was the predominantly the result of several factors: (1) heavy trimming of the mango tree, a primary producer, in the initial year of the research study, (2) a tendency for some fruit trees to produce heavily or sparingly in certain years, and (3) initiation of fruit production by recently-planted fruit trees.

Value of Food Grown

The conventional and organic values of produce that were harvested during the study period are shown in Table 27. For vegetables, the retail value of the harvested produce increased each season. The same general trend was found for fruit, with the exception of the 2013-2014 season when the mango tree produced poorly. The 3-year average retail value of harvested vegetables was $1,131 (2016 USD) for produce grown using conventional agricultural practices. Over the three-year mean, the retail organic value of the same produce was more than 70% higher than conventional agricultural pricing. This ratio between conventional and organic retail produce pricing was also found when individual prices for specific types of fruits and vegetables were compared over a growing season. Retail pricing of fruits and vegetables did not vary between stores of the same franchise but did between stores of different retailers.

The net produce value that was realized from the garden plot increased over the study period. The 3-year mean value of vegetables harvested from 1 m^2 of the garden was $16.37 (2016 USD) and was 60 percent higher when fruit values were included.

Table 25. Types of vegetables and fruits harvested from the residential food growing study over the study period.

Vegetables	Vegetables	Vegetables	Fruit
Amaranthus Arugula	Chicory (radicchio)	Pea	Banana
Basil	Chives, garlic	Snap	Coconut
Sweet	Cilantro	Snow	Grape, muscadine
Thai Holy	Delfino	Peanut	Loquat
Licorice	Flat-leaf	Pepper	Mango, kent
Purple	Collard	Bell, green	Papaya
Bean	Corn, sweet	Bell, red	Hawaiian-type
Lima	Cucumber	Serrano	Mexican-type
Soy	Daikon	Portulaca	Peach
Bean, String	Dill	Potato	Pineapple
Green pole	Eggplant	Blue	Raspberry
Green bush	Fennel	French fingerling	Strawberry
Italian flat	Garlic	Pontiac red	Watermelon
Noodle	Jicama	Yukon gold	
Royal burgundy	Kale	Potato, sweet	
Yellow wax	Kohlrabi	Beaureguard	
Bean, dry	Cossack	White	
Black	Vienna	Radish	
Pinto-type	Leek	Black globe	
Pigeon pea	Lemon grass	Red globe	
Beet	Lettuce	Rosemary	
Detroit Red	Romaine	Sage	
Golden	Bibb	Savoy	
Bok choi	Mescun	Seaweed	
White stem	Malabar	Rutabaga	
Green baby	Mint	Spinach	
Broccoli	Mushroom	Squash	
Burdock	Mustard Green	Butternut	
Cabbage	Spicy	Delicata	
Green	Nailon	Hubbard	
Purple	Nasturtium	Sprouts	
Cabbage, specialty	Onion, bulbing	Stevia	
Chinese	Red	Swiss chard	
Napa	Yellow	Thyme	
Calalu	White	Tomatillos	
Carrot	Okra	Tomato	
Danvars	Oregano	Cherry	
Purple	Parsley	Red globe type	
White	Flat leaf	Plum type	
Celery	Root	Turnip	

Table 26. Total weight of produce harvested from the residential food growing study during each growing season. Reported values exclude coconuts, which were harvested for water and not for food.

Season	Vegetables kg (lbs.)	Fruit* kg (lbs.)	Total kg (lbs.)	Vegetable Productivity per m^2	Total Productivity per m^2
2009-2010	76 (167)	3 (6)	78 (173)	1.1 (2.4)	1.0 (2.2)
2010-2011	147 (323)	104 (230)	251 (553)	1.5 (3.2)	2.3 (5.2)
2011-2012	184 (405)	201 (444)	385 (849)	2.8 (6.2)	5.3 (11.8)
2012-2013	189 (417)	419 (924)	513 (1130)	2.9 (6.4)	7.1 (15.7)
2013-2014	223 (491)	29 (65)	252 (556)	3.0 (6.6)	3.1 (6.9)
2015	n/a	941 (2074)	n/a	n/a	n/a
3-year Mean	199 (438)	463 (1021)	662 (1459)	2.9 (6.4)	8.8 (19.5)

*Lower harvest in 2009-2010 was due to severe trimming of the mango tree; increased harvest from 2010-2013 resulted from recently-planted fruit trees entering bearing age; reduced harvest during the 2013-2014 season was the result of low fruit production by the mango tree during that season.
**Growing space was defined as the area of vegetable beds plus the canopy area of fruit trees; the area of the food growing operation was more extensive because of utility areas and pathways, which covered an additional 10 sq. meters and did not change over the study period.

Agricultural census data were available from Palm Beach and Miami-Dade counties, which provided the area of land in production and the cash value of produce from those lands. Miami-Dade County is in the same region as the food growing study and, like Palm Beach County, is a significant local agricultural producer. The data for vegetables, melons and potatoes were obtained from 2012 (U.S. Agricultural Survey #1 2017). These produce types are highly comparable to the crops grown in the residential food garden study. The 2012 census reported that 37,424 hectares (92,477 acres) of these crops were harvested from Palm Beach and Miami-Dade counties with a total market value of $490,844,000 (2016 USD) or $1.31 per m^2. The average retail market value of vegetables harvested from the residential

189

Table 27. Value (2016 USD) of harvested produce from the residential food garden over the study period. Reported values exclude coconuts, which were harvested for water.

Season	Vegetables	Fruit	Total	Value Produced per m². (Vegetables / All)
2009-10				
Conventional	$452	$26	$478	$6.46 / $6.21
Organic	$734	$39	$773	$10.49 / $10.04
2010-11				
Conventional	$580	$488	$1,068	$5.80 / $9.98
Organic	$870	$727	$1,597	$8.70 / $14.93
2011-12				
Conventional	$765	$940	$1,705	$11.77 / $23.68
Organic	$1,038	$1,492	$2,530	$15.97 / $35.14
2012-13				
Conventional	$964	$1,441	$2,405	$14.83 / $34.40
Organic	$1,101	$2,229	$3,330	$16.94 / $46.25
2013-14				
Conventional	$1,665	$88	$1,753	$22.50 / $21.64
Organic	$2,590	$196	$2,786	$35.00 / $34.40
2015				
Conventional	-	$2,114	-	-
Organic	-	$3,138	-	-
3-Year mean				
Conventional	$1,131	$1498	$3,201	$16.37 / $26.57
Organic	$1,576	$2,286	$3,862	$22.64 / $38.60

All values have been converted to 2016 USD using www.usinflationcalculator.com

food growing operation was $16.37 per m² for conventional and $22.64 per m² for organic produce (Table 27). When comparing the values produced from commercial farms with those from the residential food growing operation, it is important to remember that the values reported by farms may include wholesale pricing- which is lower than what the produce was ultimately sold for in the retail market. Given this, even if the markup of farm produce was as much as 500 percent, the value of produce harvested from the residential food garden, per unit area, was more much more than that of large-scale commercial operations. These findings indicate that, per unit area, small-scale residential food gardens are much more productive and efficient food producers than large-scale commercial farms.

Production Costs

The total volume of irrigation water applied to the food garden is shown in Table 28. Under more normal climatic conditions, the garden required approximately 7,000 gallons of water per season (year). According to the United States Geological Survey (2017), the average person in the United States uses approximately 300 - 380 liters of water per day (80-100 gpd)- which would mean that the amount of irrigation water used by this food garden was comparable to an individual's water use over the period of approximately 3 months or that of a household over the period of 1-2 months.

Table 28. Total annual water usage during the case study; units are liters (gallons).

Season	Beds	Containers	Total Water Used	Well & Municipal Water Used	Captured Rainwater used (% total)	Well Pump Electricity Used (kWh)
2009-2010	11,754 (3,105)	n/a	11,754 (3,105)	11,754 (3,105)	n/a	30.8
2010-2011	63,050 (16,656)	848 (224)	63,898 (16,880)	62,637 (16,547)	333 (2)	181.1
2011-2012	35,696 (9,430)	1,798 (475)	37,495 (9,905)	27,240 (7,196)	2,709 (27)	72.4
2012-2013	16,334 (4,315)	1,650 (436)	17,984 (4,751)	5,262 (1,390)	3,361 (71)	18.1
2013-2014	24,215 (6,397)	958 (253)	25,173 (6,650)	14,014 (3,702)	2,948 (44)	110.5
3-year mean	25,415 (6,714)	1,469 (388)	26,884 (7,102)	15,505 (4,096)	3,006 (42)	67.0

Unsurprisingly, the greatest irrigation water was applied during the 2010-2011 season, which corresponded to a period of drought and extremely warm conditions. Data from this period of extreme weather conditions is useful to estimate the maximum irrigation demand that would be expected for this garden. Irrigation water was available from a rainwater collection and storage system beginning in the 2010-2011 season. The system had a storage capacity of approximately 950 liters (250 gallons) and captured water from half of the home's roof area. By increasing the capacity of the rainwater collection system by 60 percent, enough water could be acquired to meet all or most of the irrigation water needs for the garden in an average year.

The total labor expended during the case study, by task hours, is shown in Table 29. Time spent gardening, which included soil preparation, planting and plant care, was approximately 40 percent of the labor over the 3-year mean. Time spent irrigating the garden, which was primarily done by hand, was nearly half of the total time spent on growing food. This irrigation time may not be typical for most food-growing operations since hand watering was used to accurately measure water use. Installing a sprinkler or drip irrigation system would greatly reduce time spent watering.

Table 29. Time (labor costs by hour) spent on the residential food growing operation, listed by task; values shown in parenthesis indicate the percent of total time spent during the study period.

Activity	2009-2010	2010-2011	2011-2012	2012-2013	2013-2014	3-year Mean
General gardening	31.9 (46)	64.0 (28)	74.3 (47)	28.5 (35)	19.6 (27)	41 (39)
Construction/infrastructure	5.8 (8)	37.3 (16)	4.0 (3)	0.0 (0)	0 (0)	1 (1)
Composting	4.8 (7)	4.3 (2)	0.7 (0)	0.3 (0)	0 (0)	0 (0)
Manure/mulch	6.0 (9)	14.0 (6)	9.6 (6)	4.5 (5)	5.0 (7)	6 (6)
Solarizing	1.0 (1)	0.0 (0)	0.0 (0)	4.6 (6)	0.3 (0)	2 (2)
Frost protection	2.2 (3)	3.3 (1)	2.0 (1)	0.0 (0)	0.0 (0)	1 (1)
Watering*	17.3 (25)	104.0 (46)	67.9 (43)	44.5 (54)	47.3 (66)	53 (51)
Total hours worked	68.8	226.9	158.5	82.4	72.1	104

*Nearly all of the time reported in this category was for hand watering, which was necessary to measure water use and may not be typical for other operations

Operation costs of the residential food garden are shown in Table 30. These costs were broken down into different categories that included money spent on:

- Construction materials, which were usually one-time costs for permanent infrastructure such as rainwater collection system and wood for raised beds)
- Purchased soil and soil amendments (soil conditioners excluding fertilizers); purchased soil was usually potting soil used for growing in containers
- Organic fertilizers
- Organic pest control products
- Seeds, plants and other crop plant propagules
- Miscellaneous items such as signs, stakes, plant supports, etc.

The expenditures reported in Table 30 show the infrastructure investment that was required to get the food growing operation started and are summarized in the "construction materials" column. These expenditures were spread out over two years and included wood for several raised beds, materials for vertical production on the outside wall of the house (shelving units and growing containers) and rainwater collection/storage infrastructure (barrels, gutters and downspouts). Expenditures for soil purchased during the first two years of the study were higher than the remaining years, as this soil was used to fill containers for vertical production and starting seedlings. These purchases (construction materials and purchased soil) are part of the startup costs and are not annually recurring expenses.

Table 30. Monetary investment during the case study; all values converted to 2016 USD.

Season	Construction Materials	Soil & Soil Amendments	Organic Fertilizers	Organic Pest Control	Seeds & Plants	Misc.	Total
2009-2010	$378	$158	$19	$36	$11	$33	$635
2010-2011	$460	$80	$93	$45	$236	$82	$996
2011-2012	$12	$25	$22	$13	$78	$157	$307
2012-2013	$0	$27	$44	$12	$137	$13	$233
2013-2014	$0	$22	$58	$6	$47	$0	$133
3-year mean	$4	$25	$41	$10	$87	$57	$224

All values have been converted to 2016 USD using www.usinflationcalculator.com

Ecological Aspects of the Food Growing Site

Local Greenspace

A digitally processed image of the study site and its associated neighborhood is shown in Figure 18. The residential food garden, indicated on the image, was a very small portion of the total extent of greenspace in the neighborhood. This image visualizes the relatively large extent of grass within the neighborhood and shows how highly interconnected the greenspace is. It also suggests how much food growing space may be potentially available within this urban neighborhood.

Besides the relatively large extent of greenspace that could be potentially used for food growing, Figure 18 also shows how this greenspace could play an important role in the urban ecology. The combined area of urban greenspace constitutes the largest single land-cover type in cities and is a substantial proportion of a human settlement (Randall *et al.* 2003, Gaston *et al.* 2005). Urban greenspace, of which private land makes up the largest constituent (Randall *et al.* 2003), may provide the greatest contribution to urban ecology by providing multiple environmental benefits (Sperling & Lortie 2010). Urban greenspace provides important areas for groundwater recharge, water and energy conservation, and vegetation carbon sequestration (Ghosh 2010). Gardens and greenspace also mitigate the effects of daytime heat retention by paved surfaces (the urban heat island effect) (Stone & Rodgers 2001).

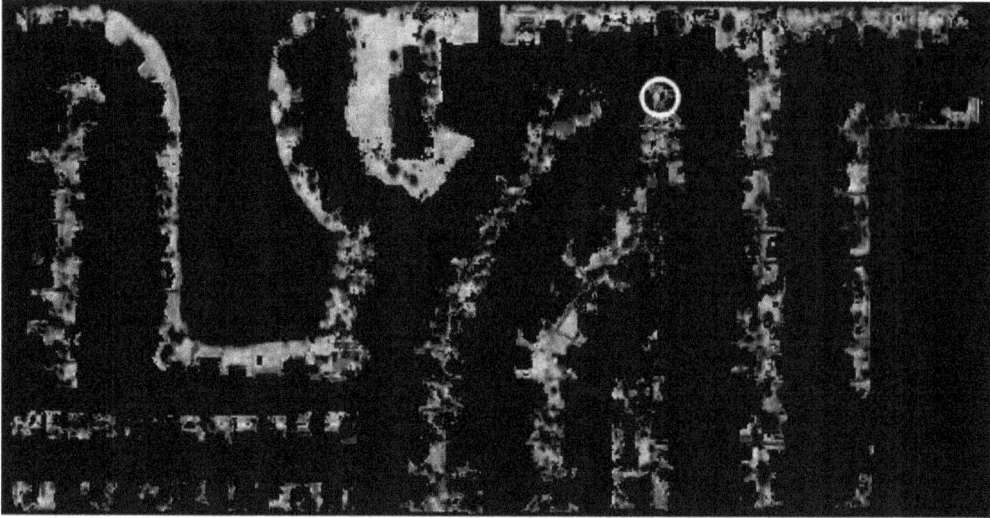

Figure 18. Digitally processed image visualizing greenspace (light areas) within the study area neighborhood (see Figure 11); the image has been digitally processed to remove trees and built structures, and to show the extent of low-profile vegetation, which is almost exclusively turfgrass. White circle indicates the location of the case study site.

Uncultivated Species Observed

A surprising number of species were observed and recorded in the residential food garden and adjacent wildlife garden. Although these observations are not intended to be a rigorous scientific survey of species present, they do offer some insight into the biodiversity potential of urban gardens in this region. These data are particularly interesting when considering that the homeowner could have decided to plant lawn instead of the food and wildlife gardens, which would have had comparatively little biodiversity and ecological function.

At least 30 species of uncultivated plants (weeds) were recorded at the study site (Table 31). These uncultivated plants included 15 species that were larval or nectar plants to 29 species of local butterflies (Minno & Minno 1999). Many of the flowering uncultivated plants attracted pollinators that are beneficial to crop production. Approximately half of the uncultivated plants were edible or produced edible fruits, many of which were common food plants in other cultures.

Table 31. Uncultivated plant species that were observed at the residential food-growing site (N=native, nn=non-native species).

Species	Ecological Significance
St. Augustine grass (*Stenotaphrum secundatum*)- N	Groundcover
Spanish needles (*Bidens alba*)- N	Nectar plant
Purslane (*Portulaca oleracea*)- nn	Nectar plant
Coffee senna (*Senna occidentalis*)- nn	Nectar plant, host plant
Smooth rattlebox (*Crotalaria pallida* var. *obovata*)- nn	Nectar plant, host plant
Panicled ticktrefoil (*Desmodium paniculatum*)-N	Nectar plant, host plant
Cabbage palm (*Sabal palmetto*)- N	Nectar plant, wildlife food plant
Carrotwood (*Cupaniopsis anacardioides*)- nn	None known
Corkystem passionflower (*Passiflora suberosa*)- N	Nectar plant, host plant, wildlife food plant
Fetid passionflower (*Passiflora foetida*)- nn	Nectar plant, host plant, wildlife food plant
Torpedo grass (*Panicum repens*)- nn	Groundcover, wildlife food plant
Lamb's-quarters (*Chenopodium album*)- nn	Wildlife food plant
Spiny amaranth (*Amaranthus spinosus*)- nn	Wildlife food plant
Clustered pellitory (*Parietaria praetermissa*)- N	None known
Leafflower (*Phyllanthus tenellus*)- nn	None known
Tropical Mexican clover (*Richardia brasiliensis*)- nn	Nectar plant
Globe sedge (*Cyperus globulosus*)- N	Wildlife food
Dog fennel (*Eupatorium capillifolium*)- N	Host plant
Ragweed (*Ambrosia artemisiifolia*)- N	Host plant
Garden spurge (*Chamaesyce hirta*)- N	Wildlife food
Spotted spurge (*Chamaesyce maculata*)- N	Wildlife food
Canada toadflax (*Linaria canadensis*)- N	Nectar plant
White-head broom (*Spermacoce verticillata*)- nn	Nectar plant
Southern sida (*Sida acuta*)- N	Nectar plant
Crowfootgrass (*Dactyloctenium aegyptium*)- nn	None known
Guinea grass (*Urochloa maxima*)- nn	None known
Balsam pear (*Momordica charantia*)- nn	Nectar plant, wildlife food plant
Unidentified Euphorbiaceae	Wildlife food
Unidentified weed 1	
Unidentified weed 2	

Numerous insect species were observed at the food growing site (Table 32) and those that were particularly well-documented (because of their size and noticeability) included butterflies and moths. Many of the butterfly species listed in Table 32 are important crop pollinators and rely on weedy or cultivated plant species at the food growing site as larval plants. The table also lists the five common insect agricultural pests that were observed during the study period.

Table 32. Invertebrate species observed at the food-growing site.

Butterflies	Observed Activity
Atala (*Eumaeus atala*)	Roosting, reproduction, nectaring
Cassius Blue (*Leptotes cassius*)	Roosting, reproduction, nectaring
Dusky Roadside Skipper (*Amblyscirtes alternate*)	Roosting, reproduction, nectaring
Eastern Black Swallowtail (*Papilio polyxenes*)	Roosting, reproduction, nectaring
Giant Swallowtail (*Papilio cresphontes*)	Roosting, nectaring
Gulf Fritillary (*Agraulis vanillae*)	Roosting, reproduction, nectaring
Hairstreak	Roosting, nectaring
Julia (*Dryas iulia*)	Roosting, reproduction, nectaring
Monarch (*Danaus plexippus*)	Roosting, reproduction, nectaring
Polydamas Swallowtail (*Battus polydamas*)	Roosting, reproduction, nectaring
Queen (*Danaus gilippus*)	Roosting, nectaring
Red Admiral (*Vanessa atalanta*)	Roosting, nectaring
Sulphur	Roosting, nectaring
Viceroy (*Limenitis archippus*)	Roosting, nectaring
White Peacock (*Anartia jatrophae*)	Nectaring
Giant Swallowtail (*Papilio cresphontes*)	Nectaring
Zebra Longwing (*Heliconius charithonia*)	Roosting, reproduction, nectaring
Unidentified moth species	
Hawk Moth spp.	Roosting, reproduction, nectaring
Tersa Sphinx Moth (*Xylophanes tersa*)	Roosting, reproduction, nectaring
Bella Moth (*Utetheisa ornatrix*)	Roosting, reproduction, nectaring
Cone-Headed Grasshopper (*Conocephalus sp.*)	Roosting, reproduction
Cricket spp.	Roosting, reproduction
Dragonfly spp.	Hunting prey
Honeybee spp.	Nectaring
Katydid spp.	Hunting prey
Ladybug spp.	Hunting prey, reproduction
Damselfly spp.	Hunting prey
Predatory Wasp spp.	Hunting prey
Other Insects	
Spotted cucumber beetle (*Diabrotica undecimpunctata howardi*)	Agricultural pest
Tomato hormworm (*Protoparce quinquemaculata*)	Agricultural pest
Fig White fly (*Singhiella simplex*)	Agricultural pest
Cucumber moth (*Diaphania indica*)	Agricultural pest
Mealy bugs	Agricultural pest

A review of the agricultural and science literature provided a list of beneficial organisms that are commonly found within the study site's region. These beneficial organisms included agricultural pest predators (Table 33), some of which target the agricultural pest insects observed in the food-growing site. Besides the agricultural pest predators, a substantial suite of pollinator insects were also observed and recorded, some of which are exclusive to some crop types that were being grown (Table 34).

Table 33. Agricultural pest predators found within the residential food garden study's region.

Resident and migratory birds	Mealybug destroyer
Ladybugs	Minute pirate bug
Praying mantis	Big eyed-bug
Soldier beetles	Assassin bug
Predatory wasps	Damsel bug
Parasitic wasps	Syrphid fly
Green lacewing	Tachinid fly

Table 34. Pollinators found within the study site's region that are associated with crop types grown in the research garden plot.

Bumblebees (14/1)	Fig wasps (1*/0)
Butterflies (1/0)	Wasps (1/0)
Carpenter bees (1/0)	Hover flies (1/1)
Honey bees (22/13)	Flies (5/4)
Leafcutter bees (1/0)	Nitilulid beetle (1*/0)
Solitary bees (15/13)	Spinx moths (1/0)
Squash bees (3/0)	Moths (1/0)
Stingless bees (5/0)	Thrips (1/0)

Number of food plant types cultivated during the study: (food/reproduction) *indicates exclusivity

Thirty-three species of birds were observed within the food growing operation site over the study period (Table 35). Of these, 52 percent were migratory species, many of which remained at the study site during the winter season. The heavy use of the study site by migratory species demonstrates the importance of urban greenspace to these species and the necessity of using growing methods that do not create risks for these organisms. Because so many migratory species spent a significant amount of time roosting and foraging within the study site, organic growing methods avoided the use of toxic materials that could poison or harm the birds that roost and forage there. The impacts of losing migratory species because of agricultural chemical use extend beyond the food growing site, even as far as northern habitats that would feel the effects of fewer returning birds.

Table 35. Vertebrate species observed at the food-growing site.

Birds	Activity	Comments
American Kestrel (*Falco sparverius*)	Roosting, foraging	Occasional visitor
American Redstart (*Setophaga ruticilla*)	Roosting, foraging	Migratory, flythrough
American White Ibis (*Eudocimus albus*)	Foraging	Permanent Resident
Baltimore Oriole (*Icterus galbula*)	Roosting, foraging	Migratory, winter resident
Black-and-White Warbler (*Mniotilta varia*)	Roosting, foraging	Migratory, winter resident
Blue Jay (*Cyanocitta cristata*)	Roosting, foraging, nesting	Permanent resident
Blue-Gray Gnatcatcher (*Polioptila caerulea*)	Roosting, foraging	Migratory, winter resident
Boat-Tailed Grackle (*Quiscalus major*)	Roosting, foraging, nesting	Permanent resident
Brown Thrasher (*Toxostoma rufum*)	Roosting, foraging	Migratory, flythrough
Carolina Wren (*Thryothorus ludovicianus*)	Roosting, foraging	Migratory, winter resident
Cedar Waxwing (*Bombycilla cedrorum*)	Roosting	Migratory, flythrough
Common Grackle (*Quiscalus quiscula*)	Roosting, foraging, nesting	Permanent resident
Common Yellowthroat (*Geothlypis trichas*)	Roosting, foraging	Migratory, winter resident
Cooper's Hawk (*Accipiter cooperii*)	Roosting, foraging	Permanent resident
Eastern Screech Owl (*Megascops asio*)	Roosting, foraging	Permanent resident
Eurasian Collared Dove (*Streptopelia decaocto*)	Roosting, foraging, nesting	Permanent resident
European Starling (*Sturnus vulgaris*)	Roosting, foraging, nesting	Permanent resident
Fish Crow (*Corvus ossifragus*)	Roosting	Migratory, winter resident
House Sparrow (*Passer domesticus*)	Roosting, foraging, nesting	Permanent resident
Monk Parakeet (*Myiopsitta monachus*)	Roosting, foraging	Permanent resident
Mourning Dove (*Zenaida macroura*)	Roosting, foraging, nesting	Permanent resident
Northern Cardinal (*Cardinalis cardinalis*)	Roosting, foraging, nesting	Permanent resident
Northern Flicker (*Colaptes auratus*)	Roosting, foraging	Occasional visitor
Northern Mockingbird (*Mimus polyglottos*)	Roosting, foraging, nesting	Permanent resident
Painted Bunting (*Passerina ciris*)	Roosting, foraging	Migratory, winter resident
Palm Warbler (*Setophaga palmarum*)	Roosting, foraging	Migratory, winter resident
Pine Warbler (*Setophaga pinus*)	Roosting, foraging	Migratory, winter resident
Prairie Warbler (*Setophaga discolor*)	Roosting, foraging	Migratory, winter resident
Red-Bellied Woodpecker (*Melanerpes carolinus*)	Roosting, foraging	Permanent resident
Red-Winged Blackbird (*Agelaius phoeniceus*)	Roosting	Migratory, flythrough
Ruby-Throated Hummingbird (*Archilochus colubris*)	Roosting, foraging	Migratory, winter resident
Spot-Breasted Oriole (*Icterus pectoralis*)	Roosting, foraging	Permanent resident
Yellow-Bellied Sapsucker (*Sphyrapicus varius*)	Roosting, foraging	Migratory; winter visitor
Yellow Rumped Warbler (*Setophaga coronata*)	Roosting, foraging	Migratory, winter resident
Other Vertebrates		
Box Turtle	Foraging	Occasional visitor
Black Snake (*Coluber constrictor*)	Foraging, nesting	Permanent resident
Brown Anole (*Anolis sagrei*)	Foraging, nesting	Permanent resident
Cane Toad (*Rhinella marina*)	Foraging	Permanent resident
Cuban Tree Frog (*Osteopilus septentrionalis*)	Foraging, nesting	Permanent resident
Glass Lizard (*Ophisaurus sp.*)	Foraging, nesting	Permanent resident
Gray Tree Frog (*Hyla versicolor*)	Foraging	Permanent resident
Knight Anole (*Anolis equestris*)	Foraging	Permanent resident
Opossum (*Didelphis virginiana*)	Foraging	Permanent resident
Raccoon (*Procyon lotor*)	Foraging	Permanent resident

Urbanization affects natural habitat in three ways- destruction (land clearing for building), degradation (impacts to remaining natural habitat) and habitat fragmentation. As the extent of urbanized land across the world continues to increase, the extent of local natural habitat is shrinking. Most urban development does not take ecological function and values into consideration, but this may have to change as the ecological goods and services that natural lands provide are being lost. Recent landscaping trends have focused on cultivating native plants, which can provide ecological function, protection of environmental quality, water savings and reduced maintenance (Diekelmann & Schuster, 2002). This is in contrast to the use of greenspace for food production, which can also provide urban sustainability benefits. Although native plant landscaping and food gardening are different in scope and purpose, they both can support sustainability issues. Urban agriculture promotes pollution awareness (Shutkin 2000, DeLind 2002) and environmentalism (Martin & Marsden 1999, Mendes *et al.* 2008). Local food production provides opportunities for recycling of materials that are in "open loop" arrangements where consumables are imported and unused portions are disposed of as waste (Smit & Nasr 1992), reducing the ecological footprint of a city (Rees 1992, Rees & Wackernagel 1996).

Private backyard gardens in cities, both food producing and non-food producing, have the capacity to enhance ecological function and connectivity (Byers 2009). One study conducted in backyard gardens in Toronto found that natural recruitment by all organisms was significant in 20 backyard study sites (Sperling & Lortie, 2010). Another study conducted in the United Kingdom found that the approximately 15 million gardens across the country functioned as bio-havens for wildlife and played a significant role in the conservation of biodiversity (Ryall & Hatherell, 2003).

Environmental Costs and Benefits

The costs and benefits described in previous sections represent the direct costs involved with the residential food growing operation. However, there were indirect costs that were also part of the operation; these are listed in Table 36. This list has two important characteristics that should be explicitly stated; first, it is comprehensive but likely not complete, as there are certainly some extended costs that are unknown. However imperfect, this table does identify the major indirect costs associated with this operation. The second important characteristic of this list is that any monetary value that could be assigned to an entry would be an estimate and not an exact value. Because of this, valuation of these extended costs carries with it imprecision and introduces error into any calculation that includes it (such as a cost-benefit analysis).

Table 36. Indirect costs of the residential food growing operation.

Material	Known Indirect Costs
Wood products (from sources that were not sustainably managed)	Forest habitat degradation associated with wood harvest. Greenhouse gas emissions associated with fossil fuel consumption from cutting and transporting wood. Environmental costs of fossil fuel extraction and use.
Garden bed borders*	Environmental impacts associated with extracting raw materials. Pollution from manufacturing processes. Environmental costs of fossil fuel extraction and use. Greenhouse gas emissions associated with transport of products.
Growing containers & vertical gardening: containers, shelving* and brackets	Environmental impacts associated with extracting raw materials. Pollution from manufacturing processes. Environmental costs of fossil fuel extraction and use. Greenhouse gas emissions associated with transport of products.
Rainwater capture and storage: PVC pipes, gutters, downspouts, rainbarrels*	Environmental impacts associated with extracting raw materials. Pollution from manufacturing processes. Environmental costs of fossil fuel extraction and use. Greenhouse gas emissions associated with transport of products.
Soil and inorganic soil amendments	Environmental impacts associated with extracting raw materials. Pollution and energy consumption associated with processing. Environmental impacts associated with manufacturing, packaging and disposal of waste (packaging) after use. Greenhouse gas emissions associated with transport of products. Environmental costs of fossil fuel extraction and use.
Organic fertilizers	Pollution and energy consumption associated with processing. Environmental impacts associated with manufacturing, packaging and disposal of waste (packaging) after use. Greenhouse gas emissions associated with transport of products. Environmental costs of fossil fuel extraction and use.
Organic pest control	Pollution and energy consumption associated with processing. Environmental impacts associated with manufacturing, packaging and disposal of waste (packaging) after use. Greenhouse gas emissions associated with transport of products. Environmental costs of fossil fuel extraction and use.
Seeds and plants	Pollution and energy consumption associated with pots, potting soil (live plants) and packaging. Greenhouse gas emissions associated with transport of products. Environmental costs of fossil fuel extraction and use.
Miscellaneous: Sheet plastic used for solarizing Frost protection* Labels Gardening tools	Environmental impacts associated with extracting raw materials. Environmental impacts associated with manufacturing, packaging and disposal of waste after use. Greenhouse gas emissions associated with transport of products. Environmental costs of fossil fuel extraction and use.
Irrigation water: Municipal or well sources	Depletion of local groundwater aquifer in an area at high risk for saltwater intrusion into the freshwater coastal aquifer.
Soil amendments: Manure, mulch and compost*	Greenhouse gas emissions associated with transport of products. Environmental costs of fossil fuel extraction and use.
Construction: Fasteners*, tools*, power usage	Environmental impacts associated with extracting raw materials. Environmental impacts associated with manufacturing, packaging and disposal of waste packaging after use. Greenhouse gas emissions associated with transport of products. Environmental costs of fossil fuel extraction and use.

*For this study, these materials came from recycled or reclaimed materials and, as such, these indirect costs were avoided

Waste Generation and Assimilation

Waste that was generated by the food growing operation could be divided into three different classes. The first class is waste products that were composted or transformed into other materials that were used as inputs to the food garden. An example of these waste products included plant material that was composted and applied to the garden as compost or mulch. A second class is recyclable materials. These included plastic, glass or metal containers that were recycled. The third class of waste products that were generated by the food growing operation required disposal and, as such, would contribute to environmental impacts. Waste products from the food growing operation are shown in Table 37.

Table 37. Non-recyclable, non-reusable waste generated by the food growing operation.

Waste Product	Fate
Plastic sheeting Approximately 37 m² used in years where solarization took place	Landfill/Waste-to-energy incinerator
Plastic bags Heavy duty plastic bags from purchased soil and soil amendments; reused for obtaining mulch and compost until no longer usable; approximately 10- 57 liter (2 cu.ft) volume bags per season	Landfill/Waste-to-energy incinerator
Irrigation system materials Remnant materials left over from installation and maintenance of irrigation system components, such as PVC pipe	Landfill/Waste-to-energy incinerator
Rainwater capture/storage system materials Remnant materials, mostly metals, left over from installation and maintenance of irrigation system components, such as scrap aluminum gutter segments	Landfill/metals recovery

Over four seasons of the study period, compostable plant-based waste material was obtained from a local food service business and the household, composted and applied as a soil amendment to the garden beds. The weight and volume of compostable material kept from the waste stream is shown in Table 38. During the 2010-2011 and 2011-2012 seasons, when compostable materials were collected throughout the entire year (compostable materials were collected for only part of the year during the other seasons), more than 1 ton of organic waste was diverted from the landfill waste stream and processed into home garden fertilizer.

For a period 7 weeks during the 2011-2012 season, the volume and weight of non-recyclable and non-compostable material that was set out for pickup and disposal to the landfill by the household was 112 kg (248 lbs.) or 0.4 kg/day (0.89 lbs./day). A comparison of the amount of compostable waste recovered (mean = 2.7

kg/day or 5.9 lbs./day) and applied to the garden with the amount of non-recyclable waste set out for landfill disposal showed a nearly 7:1 ratio of net waste assimilation by the household. This amount of net waste intake and assimilation into the food garden was limited not by garden capacity, but by the ability to acquire compostable organic material. Because of rapid degradation of soil organic matter in the south Florida climate, it was estimated that the food producing operation could have assimilated two to three times the amount of compost it assimilated over the study period.

Table 38. Organic material that was composted and added to the residential food garden (note: records were not kept during the 2013-2014 season); data are kg/m^3 (lbs./ft^3).

Season	Food Service Business	Residential Home	Total
2009-2010	220/0.7 (485/26)	29/0.1 (64/4)	249/0.8 (549/30)*
2010-2011	895/3.7 (1973/131)	104/0.3 (230/12)	999/4.0 (2203/142)
2011-2012	854/3.9 (1882/138)	92/0.3 (203/11)	946/4.2 (2085/150)
2012-2013	225/1.1 (496/37)	10/0.3 (22/11)	235/1.4 (518/48)*

*Partial data collected for these years; mean weight for years where complete data sets are available is 973 kg

Costs Associated with Energy Consumption

A summary of environmental costs related to electricity and vehicle usage during the food growing study is provided in Table 39. Over the 5-year study period, an estimated 495 kg (1,091 lbs.) of CO_2 was emitted to the atmosphere through the use of unleaded gas for transporting materials and energy use from the local utility, which was assumed to generate power using natural gas. The estimated environmental-related costs associated with use of these fossil fuels were $212.29 (2016 USD).

The produce harvested from the residential food growing operation provided a net benefit by offsetting environmental costs that would have occurred if the harvested produce had instead been purchased from a local grocery store. These benefits are shown in Table 40. The weighted mean distance (3-year mean) that food traveled from the point of origin to the grocery store was 2,676 km (1,663 miles) for conventionally grown produce and was 2,984 km (1,854 miles) for conventionally grown produce. The range of values estimated for each year compare with those that

Table 39. Environmental costs related to electricity and fossil fuel consumed during the food growing study; all currency values are 2016 USD.

	2009-2010	2010-2011	2011-2012	2012-2013	2013-2014	3-Year Mean
Electricity use						
CO_2 emissions, kg (lbs)………..	14 (31)	82 (181)	33 (73)	8 (18)	50 (110)	30 (66)
Non-greenhouse gas impacts….	$0.57	$3.36	$1.34	$0.34	$2.05	$1.24
Climate change impacts……….	$0.53	$3.15	$1.26	$0.31	$1.92	$1.16
Vehicle use						
CO_2 emissions, kg (lbs)………..	56 (123)	125 (276)	69 (152)	43 (95)	15 (33)	42 (93)
Non-greenhouse gas impacts….	$34.15	$76.06	$42.16	$25.92	$9.26	$25.78
Climate change impacts……….	$1.84	$4.10	$2.27	$1.39	$0.51	$1.39

Table 40. Climate-related impacts that would have occurred if grown produce had been purchased from local grocery stores; all currency values* are in 2016 USD.

	2009-2010	2010-2011	2011-2012	2012-2013	2013-2014	3-Year Mean
Conventionally-Grown Produce						
Weighted mean shipping distance, km (mi)	1,997 (1,241)	3,051 (1,896)	3,011 (1,871)	2,913 (1,810)	2,105 (1,308)	2,676 (1,663)
CO_2 emissions, kg (lbs)	2,482 (5,472)	3,793 (8,362)	3,741 (8,247)	3,620 (7,981)	2,615 (5,765)	3,326 (7,333)
Non-greenhouse gas impacts*	$891	$1,361	$1,342	$1,299	$938	$1,193
Climate change impacts*	$21	$32	$31	$30	$22	$28
Organically-Grown Produce						
Weighted mean shipping distance, km (mi)	3,689 (2,292)	3,716 (2,309)	3,285 (2,041)	3,067 (1,906)	2,601 (1,616)	2,984 (1,854)
CO_2 emissions, kg (lbs)	4,584 (10,106)	4,618 (10,181)	4,082 (8,999)	3,812 (8,404)	3,231 (7,123)	3,708 (8,175)
Non-greenhouse gas impacts*	$1,645	$1,657	$1,465	$1,368	$1,160	$1,331
Climate change impacts*	$38	$39	$34	$32	$27	$31

have been reported in the literature (1,500 – 2,500 miles; see Florida Fish and Game Commission & The American Farmland Trust 1995, Worldwatch Institute 2006). Organic produce had a longer transport distance than conventionally grown produce and much of it originated from California and Mexico. Over the study period, the distance that organic produce traveled decreased. This decrease is most likely the result of killing frosts that occurred in Florida during the first two years of the study, which required fresh produce to be obtained from more distant sources, and the increasing availability Florida-produced organic vegetables through time.

Data collected during this study showed that produce in the local grocery stores came from the following sources (land distances, where applicable):

- Local (80 km/50 miles)
- West Florida (200 km/125 miles)
- Georgia (800 km/500 miles)
- Virginia (1290 km/800 miles)
- Dominican Republic (1,290 km/800 miles)
- Upper Midwest of the United States (1,770 km/1,100 miles)
- Pennsylvania (1,770 km/1,100 miles)
- Texas (2,250 km/1,400 miles)
- Canada (2,740 km/1,700 miles)
- New Mexico (2,900 km/1,800 miles)
- Mexico (3,220 km/2,000 miles)
- California (4,340 km/ 2,700 miles)
- Guatemala (4,340 km/2,700 miles)
- Belize (4,830 km/3,000 miles)
- Costa Rica (5,150 km/3,200 miles)
- Colombia (6,600 km/4,100 miles)
- Holland (7,560 km/4,700 miles)
- Hawaii (8,050 km/5,000 miles)
- Peru (9,010 km/5,600 miles)
- Chile (10,460 km/6,500 miles)
- Israel (10,460 km/6,500 miles)

The estimated average (3-year mean) annual environmental-related costs associated with transporting produce would have been $1,221 for conventionally grown produce and $1331 for organically grown produce (2016 USD). These costs are based on average estimated per-mile impacts associated with heavy-duty trucks used to transport over-the-road, but the weight of the payload is not included. These costs also exclude energy-related and other environmental impacts resulting from large-scale commercial farming operations.

Intangible Costs and Benefits of Residential Food Gardens

The residential garden held importance to the household because it provided a number of intangible personal benefits, most of which could not be directly measured. These included opportunities for physical activity, the use of the garden as a place to detach from day-to-day stresses, a place for observing wildlife, connections with the residents' heritage and the garden's use as a gathering place for others in the community who wished to learn about urban food growing. Further descriptions of these and other intangible benefits to the household can be found in Zahina-Ramos 2015.

Results from the community survey of residents in the West Palm Beach metropolitan area provided insight into who grew their own food and why. Study participants who lived in households where food gardening was practiced, or had parents or grandparents who grew food, were most likely to have food gardens themselves (Figure 19). This finding indicates that familial involvement in the practice of food growing is an important part of their heritage. This is further supported by interview data, which showed that study participants' early life experiences with food gardening carried into adulthood. Respondents recounted vivid stories of relative's gardens that they visited as children, pointing to the deep connection they felt with the practice of food gardening and the loved ones who cultivated it. Those emotional connections were part of the personal drive to have a food garden and to share the experience with their children. Most interviewees expressed a strong desire to pass the practice of food growing to the next generation, hoping that their children will maintain connections to their food culture and food heritage.

Questionnaire responses related to the question of why residents grew food for themselves were divided into two classes of answers- the first class was related to the food itself and the second class was related to intangible benefits the growers received (Figure 20). Although provision of food was the most-reported reason for growing food, respondents indicated that this was to obtain vegetable and herb varieties that were not readily available from markets rather than for economic reasons. Interviewed respondents had strong feelings about how food gardening connected them to their culture, providing them with varieties of vegetables and herbs that were a part of their food heritage, and providing a sense of personal enrichment. They also felt that the quality and taste of homegrown produce was superior and it was a healthier choice than that which could be purchased from a retail store.

Food Growing in Family

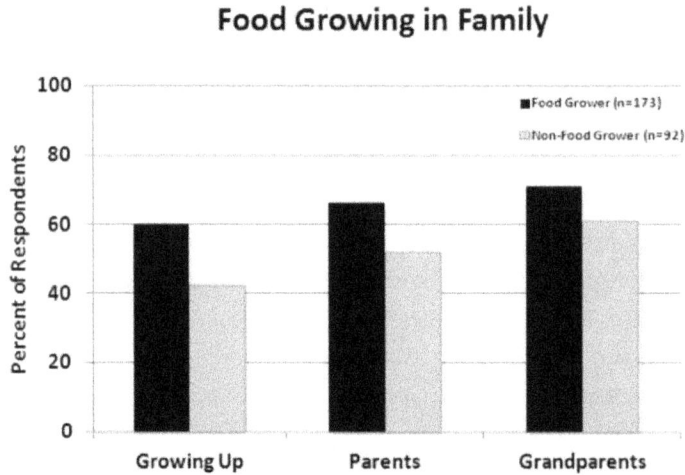

Figure 19. Association between a history of growing food within questionnaire respondents' families and the likelihood respondents were food growers; questionnaire respondents who were in a household that had a food garden (Growing Up), had parents who had a food garden when they were growing up (Parents) or who had grandparents who had food gardens (Grandparents) were all more likely to be food gardeners themselves. Figure source: Zahina-Ramos 2013.

Reasons for Food Growing

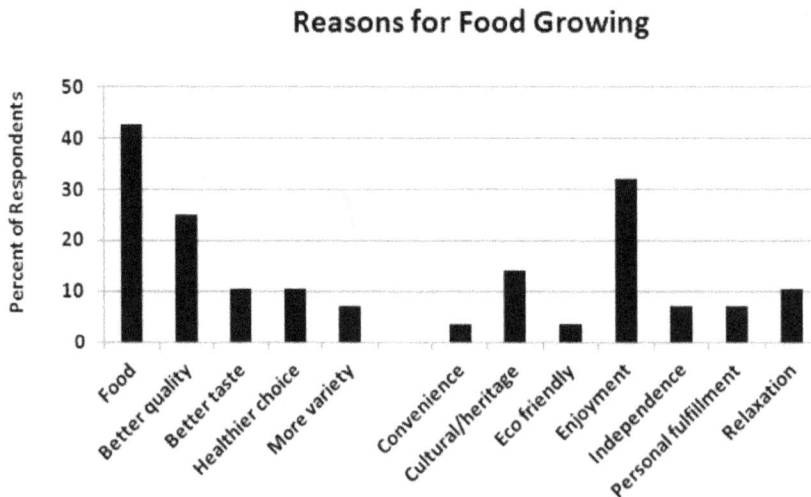

Figure 20. Questionnaire responses related to why local residents grew food for themselves; figure source: Zahina-Ramos 2013.

Questionnaire respondents reported overwhelmingly positive feelings evoked by the practice of food growing (Figure 21). Respondents reported feeling happy, relaxed, satisfied and content when food gardening. They also said that food gardening made them feel connected to their heritage, independent and appreciated.

How Food Growing Makes Respondent Feel

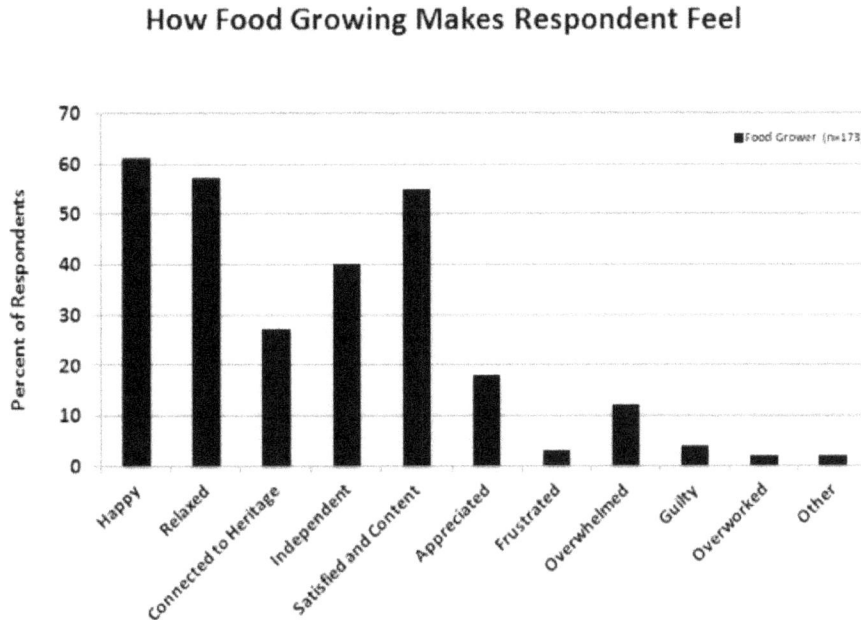

Figure 21. Intangible benefits reported from home food growers; figure source: Zahina-Ramos 2013.

Some intangible costs of the food growing operation included the feelings of being overwhelmed with the problems associated with growing food, frustration when things didn't work out, feeling overworked and feeling guilty when there was a failure they did not avoid. Results from the community survey indicated that these negative feelings were reported by a small percent of respondents, with the exception of feeling overwhelmed (12 percent). All of the reported negative feelings were associated with gardening failures (i.e. plants die, pest problems) rather than with the activity itself. Other intangible costs that were not captured in the community survey included the acceptance of risk and the burden of responsibility.

Efforts to place monetary value on the cultural or personal importance of food gardening, the feelings it evokes or the reasons why people choose to grow food for themselves will have a low probability of being accurate despite their importance. In spite of this fact, one can keep these intangible costs and benefits in mind when conducting a life-cycle assessment of a food growing operation. These intangible costs and benefits are directly associated with social sustainability concerns at the individual level.

Beyond the individual, there are also community-level costs and benefits that should be recognized. Growing food in cities contributes to household food and nutritional security (Small, 2007), and addresses some social sustainability concerns that have more recently come to the forefront. Many of these benefits have been discussed in earlier chapters of this book.

Chapter 14. Residential Food Growing Operation Case Study- Sustainability Analysis

The costs and benefits (effects) resulting from the residential food growing operation were analyzed to understand how they affect household and community sustainability. As described in a previous chapter, these effects can be placed into three groups: (1) those that can be reasonably expressed in monetary units, (2) those that can be expressed in physical units and, (3) those that can only be expressed in descriptive terms. These effects cut across the economic, environmental and social sectors of sustainability. The methods, assumptions, results and caveats associated with each analysis are presented below.

The sustainability-related effects identified from the residential food garden, which have been previously discussed, are shown in Table 41. These have been categorized by data type and reach. The following sections examine the costs and benefits of each effect in detail, which are later combined in a CBA format for analysis. When reviewing these effects, it is important to keep in mind that some of these are specific to this food growing operation and that other operations will most certainly have a different list. Also, some of the inputs to and results of the analyses are influenced by the local environmental, economic and social conditions that the study was conducted under. These will most certainly be different in other areas. For example, crop water demand in the humid climate of south Florida will certainly be different than in the arid climate of Phoenix, Arizona. As such, direct extrapolation to other regions is not recommended unless care is taken to respect these differences.

The CBAs and sustainability analyses presented below are based on result obtained from the residential food growing operation, which produced only fruits and vegetables. Humans have broader dietary needs and wants, and those considerations were not included in the food growing study. As such, the CBAs and sustainability analyses presented below do not represent the entire range of food needed for a heathy human diet. However, the methods can be applied to other agricultural forms that are part of the large food picture.

Table 41. CBA elements (effects of the food growing operation), grouped by analysis type; I = internal effects, E = external effects.

Monetary Units	Physical Units	Qualitative
Initial investment (I)	Labor inputs (I) (sometimes)	Psychological costs & benefits (I)
Fixed expenditures (I)	Food production (I)	Physical exercise (I)
Variable expenditures (I)	Food sustainability (I & E)	Food & nutrition (I & E)
Revenues (I)	Environmental costs & benefits (E)	Food knowledge benefits (I & E)
Contribution to local economy (E)	Urban ecology (I & E)	Aesthetics (I & E)
Environmental costs & benefits (E)		Community interactions & linkages (E)
Labor (I) (sometimes)		Pollution awareness (I & E)
		Agricultural support services (E)
		Food growing knowledge & skills (I & E)
		Business development (I & E)

CBA Assumptions

Labor

Because labor was a significant input to this operation, its value must be determined by some method if the net benefit from the food growing operation can be quantified. For some applications, it may be enough to report the total number of labor hours that have been expended to yield the harvested crop. This is especially true if one wishes to compare the total labor hours expended under different scenarios. This avoids the task of trying to assign value to the labor, which can be highly variable from place to place.

In other situations, it may be necessary to specify the cost or value of the labor effort expended. For food growing operations that depend on paid labor to carry out essential tasks, labor is an expense that can be calculated from paycheck receipts. For these operations, labor is a business expense. However, for food growing operations that rely on volunteer or unpaid labor (which was the case in this residential garden), no money is exchanged for labor services, so labor must be valued in another way. For these operations, labor may be valued by different methods that view the labor inputs from distinctive perspectives. A summary of these methods is provided in Chapter 10; those most applicable to this study and are locally feasible are described below.

One method that can be used to calculate the monetary value of labor is to consult with local labor pools to ascertain the average hourly rate that is paid for gardeners. A second method would be to survey unpaid workers about the per-hour value of their time; this value would be based on how much pay they would accept from an employer to do the same job. For both of these methods, the estimated hourly cost of gardeners is based on "willingness to pay" and is fixed (i.e. does not change from week to week). In addition, the value of the labor is somewhat unrelated to how much is eventually produced by the food growing operation- from the perspective of the operations manager, a certain amount of work needs to be hired for planting, maintenance and harvesting and the grower does not know how much benefit will be received by the labor investment until the harvest has been gathered.

A third method that can be used to quantifying the value of labor is to calculate the amount of revenue (food value) that the food growing operation generated. For operations that rely on unpaid labor and especially those where the workers benefit from the produced food, this may be the best approach. The reason is that labor input is not viewed as an expense but as an activity that generates revenue or that provides something that would otherwise have to be purchased, thereby increasing the amount of money retained in the workers' personal finances. Using this latter approach to labor valuation, labor costs were not included in the CBA but they are recognized as an essential input to the operation. This type of input is not monetary, but is a physical quantity that is reported. The per-hour benefit of gardening (to the residential household), according to this third method of labor valuation, is shown in Table 42.

Table 42. Labor valuation based on garden productivity; all values are 2016 USD.

	2009-2010	2010-2011	2011-2012	2012-2013	2013-2014	3-Yr Mean
Conventional Produce Prices						
Season Total	$478	$1,068	$1,705	$2,405	$1,753	$3,201
Per-hour Return for Labor	$6.95	$4.71	$10.76	$29.19	$24.31	$30.78
Per-hour Return for labor with AIS*	$8.46	$8.16	$17.83	$56.06	$58.83	$57.16
Organic Produce Prices						
Season Total	$773	$1,597	$2,530	$3,330	$2,786	$3,862
Per-hour Return for Labor	$11.24	$7.04	$15.96	$40.41	$58.83	$37.13
Per-hour Return for Labor with AIS*	$13.68	$12.20	$26.46	$77.62	$93.49	$68.96

*Estimated Per-hour Labor with AIS refers to a hypothetical scenario where an automated irrigation system (AIS) replaces the hand watering conducted during the study period, reducing the monthly labor expended for irrigation to 1 hour; see Table 29.

The food garden replaced an alternative use for the site, which was irrigated lawn for most of the neighborhood, and it is important to understand the labor that would have been spent on maintaining that landscaping. A community survey was conducted to record local resident's attitudes and perspectives about backyard food growing (Zahina-Ramos 2013); that survey also asked respondents about their (non-food producing) landscape, the time they spent on landscape maintenance and the costs involved. Of the 135 respondents that provided information about home yard maintenance, the mean time those household residents worked maintaining their landscape was 8.9 hours per month (107 hours/year). The annual labor expended to maintain home landscaping by community survey respondents is slightly higher than the average number of hours expended each year during the food growing operation (104 hours/year, see Table 29). Although the food growing operation required labor input, this must be viewed in context of what labor would have been required to maintain an alternative to the growing operation.

Labor data analysis is an important part of the CBA and its interpretation, particularly with respect to identifying ways to reduce costs and to increase benefits. Hand watering was the primary form of irrigation during the study and was necessary for accurate water use measurements. This also meant that time spent irrigating crops constituted a significant portion of the labor expended to grow food (Table 29). Table 42 shows the monetary value of food harvested, expressed as a per-hour return for labor investment. A second value, the *Estimated Per-hour Return for Labor with AIS*, is an estimate of the per-hour return that could have been realized if an automatic irrigation system (AIS) had replaced the time-consuming practice of hand watering. This analysis demonstrates how the value of labor (expressed as a function of productivity) can vary by season, can vary with respect to the harvested value that is returned and is useful to determine if a monetary investment in an AIS is financially worthwhile.

Energy Consumption and Conversion

Energy units for electricity were reported as kWh. However, the amount of fuel saved by growing locally rather than importing the food into the city from elsewhere is reported in gallons of diesel fuel. These and other issues related to quantifying energy use and savings were based on the following assumptions:

- Diesel fuel was burned at the rate of 11 km (7 miles) per gallon in the heavy-duty transport truck.
- 1kWh of electricity was equivalent to (i.e. would be obtained by burning) 0.0246 gallon of diesel fuel.
- The fuel savings realized by growing food locally rather than trucking it to the local grocery stored assumed over-the-road trucking transportation; this was the case in most instances.
- The fuel savings realized by growing food locally assumed that all of the produce grown in a single season was loaded and transported in one load, which normally does not occur.

- The fuel savings realized by growing food locally assumed that only the fuel normally consumed by the truck was burned; additional fuel would have been expended for the payload; this additional value was not included in this analysis.
- Other fossil fuel consumption associated with tilling, planting, harvesting, processing, packaging and other activities that typically occur along the commercial food supply chain are not included, but can be significant.

Environmental Costs and Benefits

Water Resources in the Local Context

The food-growing site provided a place for groundwater recharge in an urban area that contained much impervious surface. Alternative uses for this site could have been the construction of an impermeable surface (e.g. patio or other structures) that would increase urban runoff or installation of turfgrass that required irrigation, fertilizer and pesticide treatments that could pollute groundwater. Because the food growing operation was installed instead of these other alternatives, their associated environmental impacts were avoided. The net groundwater recharge to the site can be calculated, but the numbers are fraught with imprecision and requires a skilled hydrologist to analyze. Instead of quantification of the benefit, it is recognized that there is a certain quantity of groundwater recharge benefit that is realized from the food growing operation.

When compared with other alternative uses for the site, the food growing operation provided a net benefit to groundwater resources by conserving water use. According to the University of Florida's guidelines for lawn irrigation (https://edis.ifas.ufl.edu/lh025), which specify frequency and volume of lawn irrigation throughout the year, the garden footprint (101 m^2) would have required more than 55,000 gallons of water per year for irrigation if planted in lawn (assuming 50% efficiency of sprinklers and 1 inch of water applied for the mean number of water days for St. Augustine turfgrass). The garden site actually consumed, on average, 7,100 gallons of irrigation water per year under normal weather conditions- providing a net benefit of 48,000 gallons per year to the groundwater resource over irrigated lawn.

The food growing operation was situated 4.3 km (2.8 miles) from the Atlantic Ocean lagoons and coastline. Coastal communities in this area monitor coastal water supply wells for saltwater intrusion, which has become a significant risk because of high population growth, increasing demand for potable freshwater and sea level rise. Discussions with hydrogeologists at the South Florida Water Management District (West Palm Beach) indicated that during drought periods, wells in this area are at high risk for saltwater intrusion; once a water supply well becomes contaminated, it must be abandoned and a new well must be drilled. The cost for a new water supply well was estimated to be approximately $30 million (Steve Krupa, personal communication, 2013 USD), excluding planning, permitting and decommissioning the contaminated well. The food growing operation reduced the risk for saltwater intrusion into coastal wells by conserving groundwater (as compared with other potential uses for the site) in a high-risk area.

Waste Products

The food growing operation generated and assimilated waste products over the study period, some which were apparent and some that were not. Although materials were recycled when possible, others could not be. Recycled materials carried some environmental impact, which resulted from the energy and pollution that is created during the process of transforming the disposed material into a new product. It may be possible to calculate a reasonable value for this, but only when it is known what the ultimate fate of the material would be. From the perspective of resource protection, recycling materials saves resources and reduces pollution and energy consumption associated with acquiring raw materials that are being replaced by reclaimed materials. For the sake of his analysis, the amount of energy demand and pollution generated from recycling will be assumed to be less than the amount that would have resulted from using raw materials because the raw material extraction process has been bypassed. In this assumption, this residential food growing operation provided a net benefit with respect to the effects of waste creation and disposal.

Non-recycled and non-recyclable waste products from this operation were identified in Table 37 and others that were indirectly created through purchasing produces the generated waste during manufacturing or from packaging are identified in Table 36. These included:

- Waste packaging materials, mostly plastics and non-recyclable cardboard
- Plastic bags and plastic sheeting
- PVC pipe and accessories
- Remnant metals-based materials from construction

These materials were set out for garbage pickup by the local municipality and disposed of in a landfill. The costs associated with disposal, burial and decomposition of these materials can be roughly estimated. For example, the landfill charges $42 per ton (see: http://swa.org/227/Facilities-Hours) for disposal, a cost that can be calculated from the total weight of the waste that was generated by the food growing operation.

In a previous chapter, a comparison between the amount of waste that was generated by the entire household (not just the food growing operation) and the amount of waste (compostable organic material) that was assimilated by the food growing operation was discussed. Because the residential food growing operation was able take in 7 times more garbage than the household + food growing operation generated, it was a net sink for waste material in the community. The amount of energy savings and reduction of pollution associated with capturing these materials from the urban waste stream were not calculated, but could be if the appropriate experts were consulted. However, since so much more waste was consumed by the food growing operation than was being generated, a net benefit would be expected with respect to both energy savings and pollution reduction. This also helped to reduce environmental impacts associated with mining or creating essential agricultural nutrients, reducing extended impacts associated with agricultural chemical fertilizers.

Aside from the disposal fee for disposal of waste materials in a landfill, compostable organic waste produces relatively large amounts of greenhouse gasses over the long-term, particularly methane. The amount of methane that is produced by a certain weight of food waste can be calculated and added to the negative effects of disposal (see Chapter 9). Because this food growing operation captured this material from the waste stream, composted it and added it to the food garden as a soil amendment, the negative impacts of methane production by anaerobic decomposition of this material in a landfill was avoided. This amount of benefit was calculated and added to the positive effects resulting from this operation; the method for calculating this benefit was as follows:

- The amount of waste compostable material that was recovered from the waste stream over four study years are shown in Table 38; this value was for wet weight. The dry weight was assumed to be 35 percent that of the wet weight (see Chapter 12).
- For seasons where no annual data was compiled, the average daily amount of (wet) food waste that was recovered from the waste stream and composted was 2.7 kg/day (5.9 lbs./day) or 986 kg/year (2174 lbs./year). The dry weight average amount of food waste that was recovered was 345 kg/year, assuming that the dry weight was 35 percent that of the wet weight (see Chapter 12).
- Methane emissions from food waste in a landfill were estimated to be 399.5 m^3/dry Mg [refer to the US Environmental Protection Agency's WARM model (US Environmental Protection Agency 2014]. This value was used to estimate methane emissions reductions realized by capturing and composting food waste for use as a soil amendment.

- For seasons where annual average values are used, the annual amount of methane that was not created because food waste was captured and composted was 0.986 Mg x 399.5 m^3 = 393.9 m^3.

Impacts to Environmental and Natural Resources

Natural resources are extracted to create the materials and fertilizers needed to run agricultural operations. Impacts to these resources are often hidden, especially when purchasing a manufactured product that bears little resemblance to raw material it was derived from. When using lumber, it is easy to visualize that this wood came from a tree, but when using a chemical fertilizer, plastic sign or a metal post, this task becomes much more difficult. Beyond the extraction of the material, there is often environmental degradation that results from obtaining and processing raw materials. Environmental degradation can result from habitat loss and pollution associated with mining, consuming fossil fuels and transportation.

Placing monetary values on these environmental resource impacts is not possible and even extensive analyses could yield a result that has only a modest degree of accuracy. However, it is possible to quantify the amount of materials that were consumed by the food growing operation and to identify what impacts are associated with what materials. This helps to identify materials that have relatively higher or lower amount of impact to natural resources. Furthermore, by exchanging materials that have the greatest environmental impact with those that have less, the sustainability level of the food growing operation can be increased.

The materials that were used or consumed by the food growing operation carry indirect environmental costs; these are described in Table 36. Environmental impacts associated with materials used or consumed by the food growing operation are:

- Habitat degradation associated with extracting raw materials: purchased wood, growing containers for vertical gardens, brackets, hardware, soil and inorganic soil amendments.
- Greenhouse gas emissions associated with fossil fuels consumed during resource extraction and transport of materials: purchased wood, growing containers for vertical garden, brackets, hardware, rainwater capture system, soil and inorganic soil amendments.
- Pollution associated with resource refinement, manufacturing and processing: growing containers for vertical gardens, brackets, hardware, PVC for irrigation system, rainwater capture system.
- Environmental degradation associated with fossil fuel extraction and refinement: fossil fuel consumed by the above processes.

There are other costs related to fossil fuels consumed to transport food or materials that are not considered above. More directly, CO_2 emissions were produced when materials were transported for use in the food growing operation. It is assumed that for every 3.8 liters (1 gal.) of gasoline that was burned during transport, 9 kg (20 lbs.) of CO_2 were released (U.S. Department of Energy 2014). These costs are associated with the socio-political effects of fossil fuel policy and politics, which include armed conflicts, impacts to businesses caused by uncertainties caused by shifting petroleum supplies and fossil fuel market uncertainties. Although these costs were not quantified, the amount of fuel consumed or saved by the residential food growing operation were estimated from the number of miles traveled by heavy duty truck (diesel fuel) or car (unleaded gasoline).

Specific indirect environmental costs that could be identified from materials consumed or used during the food growing operation are described in Table 43. Although materials were used or consumed that had environmental impact, there were many materials that were recovered or repurposed. For example most of the garden bed borders were not purchase, but were reclaimed brick. This greatly reduced the need to purchase wood for garden borders and eliminated the need to dispose of discarded brick. Some shelving and other materials used to construct the vertical (wall) gardens, rainwater capture and retention system, protective covers for frost protection, fasteners and tools were also reclaimed or reused materials that did not add to the weight of environmental impact. It is not possible to affix monetary value to these indirect costs, but it is important to report them when listing the costs associated with the food growing operation.

Urban Ecology

The ecological benefits from the food growing operation included the creation of an approximately biodiverse greenspace within the urban environment (wildlife garden + food garden). Although it is not possible to monetarily value this space, the benefits it provides is especially important when considering that most greenspace within the neighborhood was turfgrass (Figure 18) and of limited ecological function. This greenspace was found to be used by at least 32 species of macroinvertebrates and more than 40 vertebrate species. This garden also functioned as a haven for migratory species, many of which remained in the vicinity of the site during the winter season and used it for foraging and roosting. Because organic methods were used in and around the food-growing site, these species avoided contact with harmful chemicals while utilizing the site.

Table 43. Known indirect costs that were identified from materials used or consumed during the residential food growing operation.

Material	Known Indirect Costs
Wood products (from sources that were not sustainably managed)	Forest habitat degradation associated with wood harvest. Greenhouse gas emissions associated with fossil fuel consumption from cutting and transporting wood. Environmental costs of fossil fuel extraction and use.
Garden bed borders*	Environmental impacts associated with extracting raw materials. Pollution associated with manufacturing process. Environmental costs of fossil fuel extraction and use. Greenhouse gas emissions associated with transport of products.
Growing containers & vertical gardening: containers, shelving* and brackets	Environmental impacts associated with extracting raw materials. Pollution associated with manufacturing process. Environmental costs of fossil fuel extraction and use. Greenhouse gas emissions associated with transport of products.
Rainwater capture and storage: PVC pipes, gutters, downspouts, rainbarrels*	Environmental impacts associated with extracting raw materials. Pollution associated with manufacturing process. Environmental costs of fossil fuel extraction and use. Greenhouse gas emissions associated with transport of products.
Soil and inorganic soil amendments	Environmental impacts associated with extracting raw materials. Pollution and energy consumption associated with processing. Environmental impacts associated with manufacturing, packaging and disposal of waste (packaging) after use. Greenhouse gas emissions associated with transport of products. Environmental costs of fossil fuel extraction and use.
Organic fertilizers	Pollution and energy consumption associated with processing. Environmental impacts associated with manufacturing, packaging and disposal of waste (packaging) after use. Greenhouse gas emissions associated with transport of products. Environmental costs of fossil fuel extraction and use.
Organic pest control	Pollution and energy consumption associated with processing. Environmental impacts associated with manufacturing, packaging and disposal of waste (packaging) after use. Greenhouse gas emissions associated with transport of products. Environmental costs of fossil fuel extraction and use.
Seeds and plants	Pollution and energy consumption associated with pots, potting soil (live plants) and packaging. Greenhouse gas emissions associated with transport of products. Environmental costs of fossil fuel extraction and use.
Miscellaneous: Sheet plastic used for solarizing, frost protection,* labels, and gardening tools	Environmental impacts associated with extracting raw materials. Environmental impacts associated with manufacturing, packaging and disposal of waste after use. Greenhouse gas emissions associated with transport of products. Environmental costs of fossil fuel extraction and use.
Irrigation water: Municipal or well sources	Depletion of local groundwater aquifer in an area at high risk for saltwater intrusion into the freshwater coastal aquifer.
Soil amendments: Manure, mulch and compost*	Greenhouse gas emissions associated with transport of products. Environmental costs of fossil fuel extraction and use.
Construction: Fasteners*, tools*, power usage	Environmental impacts associated with extracting raw materials. Environmental impacts associated with manufacturing, packaging and disposal of waste packaging after use. Greenhouse gas emissions associated with transport of products. Environmental costs of fossil fuel extraction and use.

*Reused, recycled or reclaimed materials

The importance of this biohaven to migratory species is especially critical with respect to birds. The southeast Florida coastline is part of the Atlantic flyway for migrating bird species, which extends from Canada to South America. One-third of the United States' population live along this flyway (National Audubon Society 2017) and the coastal area of southern Florida is among the most heavily developed along the Southeast, with much of the historic natural habitat destroyed for development. This garden provided a small, but important, contribution to conserving migrating birds. From the perspective of ecological benefit, the food-growing site contributed approximately 200 m^2 of safe and productive habitat for migrating bird species.

Examination of Costs and Benefits of the Residential Food Growing Operation

Monetary-Based Project CBA

The *monetary-based project CBA* includes only the direct expenditures and monetary benefits realized by the residential food growing operation (Figure 22), which are the economic considerations that are internal to the operation (or project). The CBA was set up for a 10-year analysis period. It was assumed that the initial-investment materials (e.g., construction materials used for raised beds and vertical growing space), which were not separately reported in the CBA but were included in the "construction materials" line item under variable costs, had a lifetime that exceeded the analysis period. The first five years of the analysis period used data collected from the residential food growing operation. The successive five years of the analysis was based on estimated costs and benefits derived from 3-year average data from the operation. Although the first five years of the analysis period were based on recorded data, summaries of the 10-year results are considered to be estimates.

Because food is a basic necessity, this monetary-based CBA considered the money saved by growing food (as compared to purchasing it) as revenue since the net result was an increase in household wealth. The annual economic benefit or loss from the food growing operation was calculated for each year for both conventional and organic food values. Since this operation employed organic growing methods, the organic food value best represents the actual benefit to the household. However, the conventional food value is reported to allow comparison between the two. The net present value (NPV) of the operation (i.e., the economic benefit to the household) over the 10-year analysis time frame was calculated for 1, 3 and 5 percent annual inflation rate scenarios (Figure 22).

INTERNAL INPUTS AND OUTPUTS

Growing Season	2009-2010	2010-2011	2011-2012	2012-2013	2013-2014	Est_FY01	Est_FY02	Est_FY03	Est_FY04	Est_FY05
Investment Costs	$0	$0	$0	$0	$0	$0	$0	$0	$0	$0
Internal Costs										
Fixed Costs	$0	$0	$0	$0	$0	$0	$0	$0	$0	$0
Variable Costs										
Labor	$0	$0	$0	$0	$0	$0	$0	$0	$0	$0
Construction materials	-$378	-$460	-$12	$0	$0	-$4	-$4	-$4	-$4	-$4
Soil amendments	-$158	-$80	-$25	-$27	-$22	-$25	-$25	-$25	-$25	-$25
Organic fertilizers	-$19	-$93	-$22	-$44	-$58	-$41	-$41	-$41	-$41	-$41
Organic pest control	-$36	-$45	-$13	-$12	-$6	-$10	-$10	-$10	-$10	-$10
Seeds, plants & propagules	-$11	-$236	-$78	-$137	-$47	-$87	-$87	-$87	-$87	-$87
Miscellaneous	-$33	-$82	-$157	-$13	$0	-$57	-$57	-$57	-$57	-$57
Electricity	-$4	-$21	-$9	-$2	-$13	-$8	-$8	-$8	-$8	-$8
Gasoline (for vehicle)	-$9	-$20	-$11	-$7	-$2	-$7	-$7	-$7	-$7	-$7
Municipal water	-$3	-$40	-$9	-$9	-$8	-$9	-$9	-$9	-$9	-$9
Total internal costs	*-$651*	*-$1,078*	*-$336*	*-$251*	*-$157*	*-$248*	*-$248*	*-$248*	*-$248*	*-$248*
Revenues										
Conventional produce prices	$478	$1,068	$1,705	$2,405	$1,753	$3,201	$3,201	$3,201	$3,201	$3,201
Organic produce prices	$773	$1,597	$2,530	$3,330	$2,786	$3,862	$3,862	$3,862	$3,862	$3,862

ANALYSIS SUMMARY

Growing Season	2009-2010	2010-2011	2011-2012	2012-2013	2013-2014	Est_FY01	Est_FY02	Est_FY03	Est_FY04	Est_FY05
Net cash flow: conv. produce price	-$173	-$10	$1,369	$2,154	$1,596	$2,953	$2,953	$2,953	$2,953	$2,953
Net cash flow: org. produce price	$122	$519	$2,194	$3,079	$2,629	$3,614	$3,614	$3,614	$3,614	$3,614

	1%	3%	5%
NPV: conventional produce prices	$18,375	$16,034	$14,051
NPV: organic produce prices	$24,911	$21,898	$19,337

Figure 22. Monetary-based (internal effects) project CBA for the residential food growing operation; all values have been placed into 2016 USD.

Expenditures over the analysis time frame included the items listed in Table 30. Line items were included in the CBA spreadsheet for startup investment costs (e.g. planning, permitting) and fixed costs (e.g., land rent, mortgage interest), but zero values were entered for these activities because there were no expenditures for these. Labor was a significant input to this operation (Table 29) and must be accounted for in the determination of net benefit, but zero values were entered into the CBA for this line item because the residential landowner carried out the gardening activity. As such, there was no expense involved with labor. However, if the operations manager had decided to hire assistance with food growing, this CBA would have been useful to determine how much cash could have been spent to cover labor cost while still generating a net profit.

The results of the monetary-based project CBA (Figure 22) showed that economic profitability was achieved the first year of the operation when considering organic pricing of produce and the second year of operation when considering conventional pricing of produce. Given the adverse growing conditions and lack of food growing experience by the gardener, these results suggest that profitability was feasible in the first year of operation under more normal weather conditions and with better preparatory training in gardening skills.

On average, urban households will spend from 10 to 40 percent of their post-tax income on home food purchases, with smaller proportions spent from higher-income households (Senauer, Asp & Kinsey 1991). The CBA showed a net financial benefit (contribution) to the residential urban household from $16,000 (conventional food pricing) to $22,000 (organic food pricing) over the 10-year analysis timeframe, assuming an inflation rate of 3 percent a year. For lower-income households, this amount of money can be an important boost to their finances. This extra money in the household budget is available to spend on other things and can be significant over longer periods of time. For example, over a 30- to 40-year period the money saved by growing a food garden of this type could be enough to purchase a modest home. The economic benefits to the household extend to the local economy as well because the money saved on food purchases was available to spend on other things within the community. This effect can be significantly multiplied when spending is directed to local businesses (Sonntag 2008). If hundreds of households in a community grew food gardens similar to the one in this study, the local economic benefit could approach millions of dollars.

Sonntag (2008) studied the role of local agricultural production in a city's economy and found that the food market economy was one major point of monetary capital leakage from the community. Sonntag concluded that local food production and local spending leads to a greater retention of cash within a community, creates more market linkages and builds stronger local food economies. By producing and purchasing from local sources, locally directed spending supports a web of local economic activity that makes for economically healthier and more prosperous communities. Spending food dollars locally can also increase regional income because the net profits are retained within the local agricultural community and local food economy businesses are likely to use local suppliers. Studies on locally directed spending indicates that this contributes as much as two to three times more to a community's income than spending at non-local businesses (Sonntag, 2008).

Comprehensive Project CBA

The CBA shown in Figure 22 only considered internal monetized effects and is the type of CBA that most businesses use when planning or analyzing their finances. However, the narrow scope of that analysis led to the exclusion of intangible effects that also play a significant role in the life of urban food growing operations. A *comprehensive project CBA* was conducted to include both tangible and intangible direct internal effects (Figure 23). This more comprehensive analysis considered the broader range of costs and benefits that resulted from the activity. Non-monetized effects were either quantified in units or qualitatively categorized as increasing benefit (desirable effect) or decreasing benefit (undesirable effect).

INTERNAL EFFECTS	2009-2010	2010-2011	2011-2012	2012-2013	2013-2014	Est_FY01	Est_FY02	Est_FY03	Est_FY04	Est_FY05
Investment Costs	$0	$0	$0	$0	$0	$0	$0	$0	$0	$0
Internal Costs										
Fixed Costs	$0	$0	$0	$0	$0	$0	$0	$0	$0	$0
Variable Costs										
Construction materials	-$378	-$460	-$12	$0	$0	-$4	-$4	-$4	-$4	-$4
Soil amendments	-$158	-$80	-$25	-$27	-$22	-$25	-$25	-$25	-$25	-$25
Organic fertilizers	-$19	-$93	-$22	-$44	-$58	-$41	-$41	-$41	-$41	-$41
Organic pest control	-$36	-$45	-$13	-$12	-$6	-$10	-$10	-$10	-$10	-$10
Seeds, plants & propagules	-$11	-$236	-$78	-$137	-$47	-$87	-$87	-$87	-$87	-$87
Miscellaneous	-$33	-$82	-$157	-$13	$0	-$57	-$57	-$57	-$57	-$57
Electricity Use	-$4	-$21	-$9	-$2	-$13	-$8	-$8	-$8	-$8	-$8
Gasoline (for vehicle)	-$9	-$20	-$11	-$7	-$2	-$7	-$7	-$7	-$7	-$7
Municipal water	-$3	-$40	-$9	-$9	-$8	-$9	-$9	-$9	-$9	-$9
Total internal costs	*-$651*	*-$1,078*	*-$336*	*-$251*	*-$157*	*-$248*	*-$248*	*-$248*	*-$248*	*-$248*
Non-Monetized Costs										
Water consumed (excluding captured rainwater) (litesr)	11,754	62,637	27,240	5,262	14,014	15,505	15,505	15,505	15,505	15,505
Labor (hrs)	68.8	226.9	158.5	82.4	72.1	104.0	104.0	104.0	104.0	104.0
Waste generated (kg)	3	3	3	3	3	3	3	3	3	3
Organic material applied to garden (kg)	249	999	946	235	973	973	973	973	973	973
Energy consumed- electricity (kWh)	30.8	181.1	72.4	18.1	110.5	67.0	67.0	67.0	67.0	67.0
Gasoline consumed (liters)	15.5	34.4	18.9	11.7	4.2	11.7	11.7	11.7	11.7	11.7
Internal Benefits										
Monetized Benefits										
Value of grown produce at conventional food prices	$478	$1,068	$1,705	$2,405	$1,753	$3,201	$3,201	$3,201	$3,201	$3,201
Value of grown produce at organic food prices	$773	$1,597	$2,530	$3,330	$2,786	$3,862	$3,862	$3,862	$3,862	$3,862
Non-Monetized Benefits										
Food produced (kg)	78	251	385	513	252	662	662	662	662	662
Fresher food and bettern nutrition	Increased	Increased	Increased	Increased	Increased	Increased	Increased	Increased	Increased	Increased
Physical exercise	Increased	Increased	Increased	Increased	Increased	Increased	Increased	Increased	Increased	Increased
Household food sustainability/security/sovereignty	Increased	Increased	Increased	Increased	Increased	Increased	Increased	Increased	Increased	Increased
Psychological benefits	Increased	Increased	Increased	Increased	Increased	Increased	Increased	Increased	Increased	Increased
Food growing knowledge and skills	Increased	Increased	Increased	Increased	Increased	Increased	Increased	Increased	Increased	Increased
Pollution awareness	Increased	Increased	Increased	Increased	Increased	Increased	Increased	Increased	Increased	Increased
Aesthetics	Increased	Increased	Increased	Increased	Increased	Increased	Increased	Increased	Increased	Increased

ANALYSIS SUMMARY

Physical Units		Non-Monetized Effects over Analysis Period	
		Qualitative	
Water consumed (excluding captured rainwater) (liters)	198,432	Fresher food and bettern nutrition	Increased
Labor (hrs)	1128.7	Physical exercise	Increased
Net waste reduction (kg)	8237	Household food sustainability/security/sovereignty	Increased
Energy consumed- electricity (kWh)	748	Psychological benefits	Increased
Gasoline consumed (liters)	143.2	Food growing knowledge and skills	Increased
Food produced (kg)	4789	Pollution awareness	Increased
		Aesthetics	Increased

Monetized Effects over Analysis Period										
Growing Season	2009-2010	2010-2011	2011-2012	2012-2013	2013-2014	Est_FY01	Est_FY02	Est_FY03	Est_FY04	Est_FY05
Net cash flow: conventional produce prices	-$173	-$10	$1,369	$2,154	$1,596	$2,953	$2,953	$2,953	$2,953	$2,953
Net cash flow: organic produce prices	$122	$519	$2,194	$3,079	$2,629	$3,614	$3,614	$3,614	$3,614	$3,614

	1%	3%	5%
NPV: conventional produce prices	$18,375	$16,034	$14,051
NPV: organic produce prices	$24,911	$21,898	$19,337

Figure 23. Comprehensive project CBA for the residential food growing operation.

The comprehensive project CBA included quantification of the water consumed by the operation (other than captured rainwater), which is important to understand how this food growing activity may impact limited local water resources. It also lists the number of hours worked, the amount of (non-recycled) waste material that was generated, the amount of organic waste material (mostly food scraps) that were composted and applied to the garden, energy consumed to run the well pump for irrigation, and the amount of gasoline that was consumed to transport materials during the study period. Although these effects were reported in physical units rather than dollars, they are useful to understand the magnitude of these effects as well as for comparison between alternative scenarios.

Other effects represented in the comprehensive project CBA were those that were not quantifiable, but were important nonetheless. These included access to better-quality food, opportunities for physical exercise, contribution to the household's food security and other effects. As with the effects that were reported as physical units, identifying these is important to understand the fuller impact of the food growing operation and to permit comparisons with other alternative scenarios.

Results from the comprehensive project CBA showed no change in the net financial benefit to the household as compared to the monetary-based CBA. This food growing operation would consume nearly 200,000 liters (52,830 gal.) of water over the analysis period, which constituted a small draw on the region's water resources. The food growing operation would also generate 30 kg of waste material. When compared with the amount of organic material that was composted and applied as a soil amendment, there would be a net intake of 8237 kg (9.0 tons) of waste material over the analysis period. The composed organic material was free of charge (although labor and fuel for transportation were required) but if it had to be purchased, this additional cost would affect the economic profitability of the operation. This analysis is a valuable tool for determining how much money could be expended for organic soil amendments while maintaining profitability. Intangible benefits that were included in the analysis made positive contributions to the well being of the household that lie beyond the realm of finances.

To avoid the perception of double counting, it is important to recognize that the food grown by this operation was represented twice in this CBA because it produced two different types of benefit. The first benefit originated from the positive financial effect realized because the household had to purchase less food to meet food needs. This benefit is closely tied to the labor hours expended. A second type of benefit realized from the grown food was the household's access to 4789 kg (5.2 tons) of produce (over the analysis period), which had dietary and food security implications.

Comprehensive Efficiency CBA

The comprehensive efficiency CBA includes internal and external effects that are both monetized and non-monetized (Figure 24). As with the comprehensive project CBA, non-monetized effects are presented as physical units or are qualitatively described. This CBA included the following external effects as monetary units:

- Greenhouse and non-greenhouse gas impacts from energy consumption, which was a cost of the operation (see Chapter 9).

INTERNAL AND EXTERNAL EFFECTS	2009-2010	2010-2011	2011-2012	2012-2013	2013-2014	Est_FY01	Est_FY02	Est_FY03	Est_FY04	Est_FY05
Investment Costs	$0	$0	$0	$0	$0	$0	$0	$0	$0	$0
Internal Costs										
Fixed Costs	$0	$0	$0	$0	$0	$0	$0	$0	$0	$0
Variable Costs										
Construction materials	-$378	-$460	-$12	$0	$0	-$4	-$4	-$4	-$4	-$4
Soil amendments	-$158	-$80	-$25	-$27	-$22	-$25	-$25	-$25	-$25	-$25
Organic fertilizers	-$19	-$93	-$22	-$44	-$58	-$41	-$41	-$41	-$41	-$41
Organic pest control	-$36	-$45	-$13	-$12	-$6	-$10	-$10	-$10	-$10	-$10
Seeds, plants & propagules	-$11	-$236	-$78	-$137	-$47	-$87	-$87	-$87	-$87	-$87
Miscellaneous	-$33	-$82	-$157	-$13	$0	-$57	-$57	-$57	-$57	-$57
Electricity Use	-$4	-$21	-$9	-$2	-$13	-$8	-$8	-$8	-$8	-$8
Gasoline (for vehicle)	-$9	-$20	-$11	-$7	-$2	-$7	-$7	-$7	-$7	-$7
Municipal water	-$3	-$40	-$9	-$9	-$8	-$9	-$9	-$9	-$9	-$9
Total internal costs	-$651	-$1,078	-$336	-$251	-$157	-$248	-$248	-$248	-$248	-$248
Non-Monetary Costs										
Water consumed (excluding captured rainwater) (liter)	11,754	62,637	27,240	5,262	14,014	15,505	15,505	15,505	15,505	15,505
Labor (hrs)	68.8	226.9	158.5	82.4	72.1	104.0	104.0	104.0	104.0	104.0
Waste generated (kg)	3	3	3	3	3	3	3	3	3	3
Organic material applied to garden (kg)	249	999	946	235	973	973	973	973	973	973
Energy consumed- electricity (kWh)	30.8	181.1	72.4	18.1	110.5	67.0	67.0	67.0	67.0	67.0
Gasoline consumed (liters)	15.5	34.4	18.9	11.7	4.2	11.7	11.7	11.7	11.7	11.7
Internal Benefits										
Monetary Benefits										
Value of grown produce at conventional food prices	$478	$1,068	$1,705	$2,405	$1,753	$3,201	$3,201	$3,201	$3,201	$3,201
Value of grown produce at organic food prices	$773	$1,597	$2,530	$3,330	$2,786	$3,862	$3,862	$3,862	$3,862	$3,862
Non-Monetary Benefits										
Food produced (kg)	78	251	385	513	252	662	662	662	662	662
Fresher food and betterm nutrition	Increased	Increased	Increased	Increased	Increased	Increased	Increased	Increased	Increased	Increased
Physical exercise	Increased	Increased	Increased	Increased	Increased	Increased	Increased	Increased	Increased	Increased
Household food sustainability/security/sovereignty	Increased	Increased	Increased	Increased	Increased	Increased	Increased	Increased	Increased	Increased
Psychological benefits	Increased	Increased	Increased	Increased	Increased	Increased	Increased	Increased	Increased	Increased
Food growing knowledge and skills	Increased	Increased	Increased	Increased	Increased	Increased	Increased	Increased	Increased	Increased
Pollution awareness	Increased	Increased	Increased	Increased	Increased	Increased	Increased	Increased	Increased	Increased
Aesthetics	Increased	Increased	Increased	Increased	Increased	Increased	Increased	Increased	Increased	Increased
External Costs										
Monetary Costs										
Non Greenhouse Gas Impacts from electricity Use	-$1	-$3	-$1	$0	-$2	-$1	-$1	-$1	-$1	-$1
Climate Change Impacts from electricity Use	-$1	-$3	-$1	$0	-$2	-$1	-$1	-$1	-$1	-$1
Non Greenhouse Gas Impacts from vehicle Use	-$34	-$76	-$42	-$26	-$9	-$26	-$26	-$26	-$26	-$26
Climate Change Impacts from vehicle Use	-$2	-$4	-$2	-$1	-$1	-$1	-$1	-$1	-$1	-$1
Total external monetary costs	-$37	-$87	-$47	-$28	-$14	-$30	-$30	-$30	-$30	-$30
Non-Monetary Costs										
CO_2 emissions to atmosphere from electicity use (kg)	14	82	33	8	50	30	30	30	30	30
CO_2 emissions to from vehicle use (kg)	56	125	69	43	15	42	42	42	42	42
Saltwater intrusion risk to potable groundwater resources	Increased	Increased	Increased	Increased	Increased	Increased	Increased	Increased	Increased	Increased
External Benefits										
Monetary Benefits										
Avoided environmental impacts of chemical fertilizers	$7.69	$10.50	$7.45	$7.45	$8.22	$7.69	$7.69	$7.69	$7.69	$7.69
Avoided disposal fee for food waste that was composted	$21	$42	$42	$21	$42	$42	$42	$42	$42	$42
Fossil fuel impacts from long-term produce transport- conv. grown produce										
Avoided non Greenhouse Gas Impacts	$891	$1,361	$1,342	$1,299	$938	$1,193	$1,193	$1,193	$1,193	$1,193
Avoided Greenhouse Gas Impacts	$21	$32	$31	$30	$22	$28	$28	$28	$28	$28
Fossil fuel impacts from long-term produce transport- org. grown produce										
Avoided non Greenhouse Gas Impacts	$1,645	$1,657	$1,465	$1,368	$1,160	$1,331	$1,331	$1,331	$1,331	$1,331
Avoided Greenhouse Gas Impacts	$38	$39	$34	$32	$27	$31	$31	$31	$31	$31
Non-Monetary Benefits										
Provided area beneficial to urban ecology (m^2)	101	138	98	98	108	101	101	101	101	101
Avoided CO_2 emissions from transporting purchased conv.-grown produce (kg)	2482	3793	3741	3620	2615	3326	3326	3326	3326	3326
Avoided CO_2 emissions from transporting purchased org.-grown produce (kg)	4584	4618	4082	3812	3231	3708	3708	3708	3708	3708
Diesel fuel not consumed for transport because food was grown locally										
Convention produce (liters)	670	1026	1011	980	708	901	901	901	901	901
Organic produce (liters)	1238	1249	1105	1030	874	1003	1003	1003	1003	1003
Methane emissions avoided by capturing/composting food waste (m3)	99.5	399.1	377.9	93.9	393.9	393.9	393.9	393.9	393.9	393.9
Avoided environmental destruction associated with phosphate mining	Increased	Increased	Increased	Increased	Increased	Increased	Increased	Increased	Increased	Increased
Community interaction and linkages	Increased	Increased	Increased	Increased	Increased	Increased	Increased	Increased	Increased	Increased
Community food sustainability	Increased	Increased	Increased	Increased	Increased	Increased	Increased	Increased	Increased	Increased

ANALYSIS SUMMARY

Non-Monetized Effects over Analysis Period

Physical Units		Qualitative	
Water consumed (excluding captured rainwater) (liters)	198,432	Saltwater intrusion risk to potable groundwater resources	Increased
Labor (hrs)	1,129	Avoided environmental destruction associated with phosphate mining	Increased
Net waste produced (kg)	-8,237	Fresher food and betterm nutrition	Increased
Energy consumed- electricity (kWh)	748	Physical exercise	Increased
Gasoline consumed (liters)	143.2	Household food sustainability/security/sovereignty	Increased
Food produced (kg)	4,789	Psychological benefits	Increased
Provided area beneficial to urban ecology (mean m^2)	105	Food growing knowledge and skills	Increased
Avoided methane emissions by capturing/composting food waste (m3)	3,334	Pollution awareness	Increased
		Aesthetics	Increased
Avoided CO_2 emissions by growing food locally		Community interaction and linkages	Increased
Conventionally grown produce (kg)	32,026	Community food sustainability	Increased
Organically grown produce (kg)	38,012		
Diesel fuel not consumed for transport because food was grown locally			
Convention produce (liters)	8,900		
Organic produce (liters)	10,511		

Monetized Effects over Analysis Period

Growing Season	2009-2010	2010-2011	2011-2012	2012-2013	2013-2014	Est_FY01	Est_FY02	Est_FY03	Est_FY04	Est_FY05
Net cash flow: conventional produce prices	$731	$1,349	$2,745	$3,484	$2,593	$4,194	$4,194	$4,194	$4,194	$4,194
Net cash flow: organic produce prices	$1,797	$2,181	$3,696	$4,480	$3,853	$4,996	$4,996	$4,996	$4,996	$4,996

	1%	3%	5%
NPV: conventional produce prices	$29,894	$26,395	$23,417
NPV: organic produce prices	$38,547	$34,224	$30,535

Figure 24. Comprehensive Efficiency CBA for the residential food growing operation.

- Greenhouse and non-greenhouse gas impacts that were avoided because food was grown locally rather than purchased from a grocery store. Since most produce offered by grocery stores had been grown elsewhere and transported, this effect was a savings (not a cost) resulting from the operation. Because organically grown produce traveled, on average, a greater distance, the amount of fossil fuel consumed during transport was higher.
- Environmental impacts related to chemical fertilizer use; because chemical fertilizers were not used, this was a savings rather than a cost (see Chapter 9).
- Because the food growing operation took in much more compostable organic waste than it generated, it was a net sink. The material would have gone to the local landfill. Because the food growing operation avoided this cost, it was included as a savings.

Non-monetized effects that were expressed in quantitative units were:
- The amount of CO_2 emissions that were created from energy consumption.
- The amount of CO_2 emissions that were avoided because food was grown locally rather than purchased from a grocery store. Since most produce offered by grocery stores had been grown elsewhere and transported, this effect was a savings (not a cost) resulting from the operation.
- The amount of fuel that was not consumed because food was grown locally rather than purchased from a grocery store. Since most produce offered by grocery stores had been grown elsewhere and transported, this effect was a savings (not a cost) resulting from the operation.
- The area of the garden that provided safe habitat for resident and migratory organisms.
- The amount of methane emissions that did not occur because compostable organic waste was not buried in a landfill.

Other effects of the food growing operation that were reported as qualitative descriptions were the impact of the operation on saltwater intrusion risk to local groundwater resources and the operation's influence on community food sustainability and interactions.

As compared with the project CBAs, the monetary value of the operation increased in the efficiency CBA more than 50 percent, mostly due to the inclusion of avoided impacts from fossil fuel consumption. Other important savings identified by the analysis included a net intake of waste material from the community, a reduction in methane emissions by 3334 m^3, a reduction in CO_2 emissions by 32,000 - 38,000 kg (35 - 42 tons) and a savings of up to 10,000 liters (2,600 gal.) of diesel fuel over the analysis period. In addition to the qualitative benefits identified in the comprehensive project CBA, additional community benefits were identified with respect to increasing community interactions/linkages and food sustainability. However, this operation did contribute to increasing the risk for saltwater intrusion into potable groundwater resources, but the effect was small- comparable to the amount of water consumed by the average American household over several months. The results from the comprehensive efficiency CBA illustrates why it is important to consider both internal and external effects when examining how a project or activity impacts a community.

With-Project/Without-Project Analysis

Although the above CBAs described how the residential food garden benefitted the household and community, an analysis of a with-project/without-project scenario is beneficial to understand the larger impact the residential food garden had because it occurred. This analysis is different in scope than the previous CBA analyses because it considers the effects that occurred because the project replaced an alternative use for the site. For this analysis, it will be assumed that the household would have installed lawn (the alternative use for the site) instead of the food growing operation. This lawn would have been similar to others in the neighborhood, it would have been managed according to the University of Florida guidelines for St. Augustine turfgrass (the standard lawn grass in the region) and it would have been fertilized regularly. Time spent maintaining the lawn would have been by the homeowner, rather than by a hired landscape maintenance company. It was assumed that each month, the homeowner would have spent 8.9 hours (on average) on maintenance; this estimate came from the community survey conducted by Zahina-Ramos (2013). Monthly costs for lawn maintenance supplies, including fertilizers and pesticides, was assumed to be $20/month plus gas costs ($0.60 per liter or $2.23 per gallon) for the lawn mower. The amount of gasoline consumed for lawn maintenance was derived from estimates from the University of Vermont Extension (2016) and pro-rated for the food growing area.

Additional effects that were included in the With/Without Project scenario were:

- Opportunity cost- using the site for food growing prevented use of the site for other activities, such recreation
- Avoided the labor associated with lawn and landscape maintenance

- Avoided the costs for lawn fertilizers, pest control, mower and associated materials
- Avoided burning fossil fuels for lawn maintenance
- The reduction in greenhouse gasses that were created because fossil fuels were not consumed by lawn maintenance activities
- A reduction in impacts related to above greenhouse gasses
- Water savings resulting from the decision to grow a food garden in the space rather than installing irrigated lawn
- A reduction in risk for saltwater intrusion into the local coastal aquifer system related to less irrigation water use

The With/Without Project CBA is shown in Figure 25. The monetary benefit arising from the food growing operation increased because the expense of maintaining a lawn was avoided. Most notably, over the 10-year analysis period, the amount of time spent to do the food garden was only 59 hours (5.9 hours per year) more than would have been expended if the area had been planted in lawn. The decision to create a food garden offered the household with an abundance of food whereas the lawn would have provided little benefit beyond aesthetics.

Other benefits realized by the decision to plant a food garden include a net reduction in the risk for saltwater intrusion into the local groundwater aquifer. This is because the food garden used approximately 1,420,000 liters (375,000 gal.) less water than the comparable area of irrigated lawn over the analysis period and represents a significant savings to the resource. Other benefits were realized in the areas of energy conservation and a reduction in greenhouse gasses.

Sensitivity Analysis and Uncertainty Management

It is recognized that these CBAs contain assumptions and effects that are not shared by all food growing operations. They also contain estimates that are of limited accuracy. However, these examples do demonstrate how the CBA can be applied to urban agricultural systems to determine how they can benefit or negatively impact the operator or community. The CBA, when properly constructed and interpreted, is a valuable decision-making tool for operations managers, planners, policy makers and politicians who must determine how much benefit can be derived from investment in sustainable urban agricultural systems.

There are certainly omissions and imprecise data included in these CBAs- but that does not detract from their usefulness. These CBAs identify the major effects and quantifies them in reasonable terms. There is no guarantee or assurance of absolute accuracy (no such analysis can provide that), however they do provide a reasonable estimate of the net costs and benefits.

WITH/WITHOUT PROJECT EFFECTS	2009-2010	2010-2011	2011-2012	2012-2013	2013-2014	Est_FY01	Est_FY02	Est_FY03	Est_FY04	Est_FY05
Investment Costs	$0	$0	$0	$0	$0	$0	$0	$0	$0	$0
Internal Costs										
Fixed Costs	$0	$0	$0	$0	$0	$0	$0	$0	$0	$0
Variable Costs										
Construction materials	-$378	-$460	-$12	$0	$0	-$4	-$4	-$4	-$4	-$4
Soil amendments	-$158	-$580	-$25	-$27	-$22	-$25	-$25	-$25	-$25	-$25
Organic fertilizers	-$19	-$93	-$22	-$44	-$58	-$41	-$41	-$41	-$41	-$41
Organic pest control	-$36	-$45	-$13	-$12	-$6	-$10	-$10	-$10	-$10	-$10
Seeds, plants & propagules	-$11	-$236	-$578	-$137	-$47	-$87	-$87	-$87	-$87	-$87
Miscellaneous	-$33	-$82	-$157	-$13	$0	-$57	-$57	-$57	-$57	-$57
Electricity Use	-$4	-$21	-$9	-$2	-$13	-$8	-$8	-$8	-$8	-$8
Gasoline (for vehicle)	-$9	-$20	-$11	-$7	-$2	-$7	-$7	-$7	-$7	-$7
Municipal water	-$3	-$40	-$9	-$9	-$8	-$9	-$9	-$9	-$9	-$9
Total Internal costs	-$651	-$1,078	-$336	-$251	-$157	-$248	-$248	-$248	-$248	-$248
Non-Monetary Costs										
Water consumed (excluding captured rainwater) (liter)	11,754	62,637	27,240	5,262	14,014	15,505	15,505	15,505	15,505	15,505
Labor (hrs)	68.8	226.9	158.5	82.4	72.1	104.0	104.0	104.0	104.0	104.0
Waste generated (kg)	3	3	3	3	3	3	3	3	1	3
Waste organic material applied to garden (kg)	249	999	946	235	973	973	973	973	973	973
Energy consumed- electricity (kWh)	30.8	181.1	72.4	18.1	110.5	67.0	67.0	67.0	67.0	67.0
Gasoline consumed (liters)	15.5	34.4	18.9	11.7	4.2	11.7	11.7	11.7	11.7	11.7
Internal Benefits										
Monetary Benefits										
Value of grown produce at conventional food prices	$478	$1,068	$1,705	$2,405	$1,753	$3,201	$3,201	$3,201	$3,201	$3,201
Value of grown produce at organic food prices	$773	$1,597	$2,530	$3,330	$2,786	$3,862	$3,862	$3,862	$3,862	$3,862
Non-Monetary Benefits										
Food produced (kg)	78	251	385	513	252	662	662	662	662	662
Fresher food and better nutrition	Increased	Increased	Increased	Increased	Increased	Increased	Increased	Increased	Increased	Increased
Physical exercise	Increased	Increased	Increased	Increased	Increased	Increased	Increased	Increased	Increased	Increased
Household food sustainability/security/sovereignty	Increased	Increased	Increased	Increased	Increased	Increased	Increased	Increased	Increased	Increased
Psychological benefits	Increased	Increased	Increased	Increased	Increased	Increased	Increased	Increased	Increased	Increased
Food growing knowledge and skills	Increased	Increased	Increased	Increased	Increased	Increased	Increased	Increased	Increased	Increased
Pollution awareness	Increased	Increased	Increased	Increased	Increased	Increased	Increased	Increased	Increased	Increased
Aesthetics	Increased	Increased	Increased	Increased	Increased	Increased	Increased	Increased	Increased	Increased
External Costs										
Monetary Costs										
Non Greenhouse Gas impacts from electricity Use	-$1	-$3	-$1	$0	-$2	-$1	-$1	-$1	-$1	-$1
Climate Change impacts from electricity Use	-$1	-$3	-$1	$0	-$2	-$1	-$1	-$1	-$1	-$1
Non Greenhouse Gas impacts from vehicle Use	-$34	-$76	-$42	-$26	-$9	-$26	-$26	-$26	-$26	-$26
Climate Change impacts from vehicle Use	-$2	-$4	-$2	-$1	-$1	-$1	-$1	-$1	-$1	-$1
Total external monetary costs	-$37	-$87	-$47	-$28	-$14	-$30	-$30	-$30	-$30	-$30
Non-Monetary Costs										
CO_2 emissions to atmosphere from electricity use (kg)	14	82	33	8	50	30	30	30	30	30
CO_2 emissions to from vehicle use (kg)	56	125	69	43	15	42	42	42	42	42
Saltwater intrusion risk to potable groundwater resources	Increased	Increased	Increased	Increased	Increased	Increased	Increased	Increased	Increased	Increased
External Benefits										
Monetary Benefits										
Avoided environmental impacts of chemical fertilizers	$7.69	$10.50	$7.45	$7.45	$8.22	$7.69	$7.69	$7.69	$7.69	$7.69
Avoided disposal fee for food waste that was composted	$21	$42	$42	$21	$42	$42	$42	$42	$42	$42
Fossil fuel impacts from long-term produce transport- conv. grown produce										
Avoided non Greenhouse Gas impacts	$891	$1,361	$1,342	$1,299	$938	$1,193	$1,193	$1,193	$1,193	$1,193
Avoided Greenhouse Gas impacts	$21	$32	$31	$30	$22	$28	$28	$28	$28	$28
Fossil fuel impacts from long-term produce transport- org. grown produce										
Avoided non Greenhouse Gas impacts	$1,645	$1,657	$1,465	$1,368	$1,160	$1,331	$1,331	$1,331	$1,331	$1,331
Avoided Greenhouse Gas impacts	$38	$39	$34	$32	$27	$31	$31	$31	$31	$31
Non-Monetary Benefits										
Provided area beneficial to urban ecology (m^2)	101	138	98	98	108	101	101	101	101	101
Avoided CO_2 emissions from transporting purchased conv. grown produce (kg)	2482	3793	3741	3620	2615	3326	3326	3326	3326	3326
Avoided CO_2 emissions from transporting purchased org. grown produce (kg)	4584	4618	4082	3812	3231	3708	3708	3708	3708	3708
Diesel fuel not consumed for transport because food was grown locally										
Convention produce (liters)	670	1026	1011	980	708	901	901	901	901	901
Organic produce (liters)	1238	1249	1105	1030	874	1003	1003	1003	1003	1003
Methane emissions avoided by capturing/composting food waste (m3)	99.5	399.1	377.9	93.9	393.9	393.9	393.9	393.9	393.9	393.9
Avoided environmental destruction associated with phosphate mining	Increased	Increased	Increased	Increased	Increased	Increased	Increased	Increased	Increased	Increased
Community interest on and linkages	Increased	Increased	Increased	Increased	Increased	Increased	Increased	Increased	Increased	Increased
Community food sustainability	Increased	Increased	Increased	Increased	Increased	Increased	Increased	Increased	Increased	Increased
With-Without Project										
With Project Costs										
Availability of space for other activies	Decreased	Decreased	Decreased	Decreased	Decreased	Decreased	Decreased	Decreased	Decreased	Decreased
With Project Benefits										
Avoided CO_2 emissions from landscape maintenance (kg)	11.9	16.4	11.5	11.5	12.6	11.9	11.9	11.9	11.9	11.9
Avoided greenhouse gas impacts from fuel consumed for landscape maintenance	$0.53	$0.53	$0.37	$0.37	$0.41	$0.39	$0.39	$0.39	$0.39	$0.39
Avoided fertilizer, pest control and other maintenance costs	$251	$256	$251	$251	$252	$251	$251	$251	$251	$251
Avoided consuming fuel (gasoline) for landscape maintenance (liters)	5.1	7.0	4.9	4.9	5.4	5.1	5.1	5.1	5.1	5.1
Avoided landscape maintenance (hours)	107	107	107	107	107	107	107	107	107	107
Landscape irrigation water not needed (liters)	169,946	119,063	154,460	176,488	167,686	166,195	166,195	166,195	166,195	166,195
Saltwater intrusion risk to local groundwater resources	decrease	decrease	decrease	decrease	decrease	decrease	decrease	decrease	decrease	decrease

ANALYSIS SUMMARY

Non-Monetized Effects over Analysis Period			
Physical Units		**Qualitative**	
Water consumed (excluding captured rainwater) (liters)	-1,420,136	Saltwater intrusion risk to potable groundwater resources	Increased
Labor (hrs)	59	Avoided environmental destruction associated with phosphate mining	Increased
Net waste produced (kg)	-8,237	Fresher food and better nutrition	Increased
Energy consumed- electricity (kWh)	748	Physical exercise	Increased
Gasoline consumed (liters)	90	Household food sustainability/security/sovereignty	Increased
Food produced (kg)	4,789	Psychological benefits	Increased
Provided area beneficial to urban ecology (mean m^2)	105	Food growing knowledge and skills	Increased
Avoided methane emissions by capturing/composting food waste (m3)	3,334	Pollution awareness	Increased
		Aesthetics	Increased
Avoided CO_2 emissions by growing food locally		Community interaction and linkages	Increased
Conventionally grown produce (kg)	32,150	Community food sustainability	Increased
Organically grown produce (kg)	38,136	Saltwater intrusion risk to potable groundwater resources	Decreased
		Availability of space for other activies	Decreased
Diesel fuel not consumed for transport because food was grown locally			
Convention produce (liters)	8,900		
Organic produce (liters)	10,911		

Monetized Effects over Analysis Period									
Growing Season	2009-2010	2010-2011	2011-2012	2012-2013	2013-2014	Est_FY01	Est_FY02	Est_FY03	Est_FY04	Est_FY05
Net cash flow: conventional produce prices	$983	$1,605	$2,996	$3,735	$2,846	$4,446	$4,446	$4,446	$4,446	$4,446
Net cash flow: organic produce prices	$2,049	$2,437	$3,947	$4,731	$4,106	$5,248	$5,248	$5,248	$5,248	$5,248
	1%	3%	5%							
NPV: conventional produce prices	$32,283	$28,547	$29,365							
NPV: organic produce prices	$40,937	$36,377	$32,484							

Figure 25. With-Project CBA for the residential food growing operation.

A sensitivity analysis can be conducted to determine the effects that contribute most to the net result. This can be done with both the monetary and physical unit effects. In the CBA examples provide above, estimates for the monetary cost of damage from climate change were included in the analysis. If the analyst has reason to believe that the accuracy of the assumed values may be +/- 25 percent, then values that are 25 percent higher and lower than the assumed value may be placed into the analysis to see how much these may change the outcome. A similar approach can be used with the physical unit effects. The sensitivity analysis can be used to identify and manage uncertainty in the operation and its data.

A sensitivity analysis may also be conducted to understand where thresholds may exist in the operation. For example, by how much does fossil fuel use need to be reduced to obtain a net-zero greenhouse gas emissions scenario? How closely can the operation come to reaching this goal if electricity consumption were replaced with on-site solar energy generation? How much can be spent on materials before the operation ceases to be financially profitable? These and other questions can be examined by a properly crafted spreadsheet CBA.

Residential Food Production and Food Sustainability

The food sustainability metrics presented in Chapter 11 were calculated based on data from the residential food growing operation study. Although the results of this analysis are interesting and informative, they are also valuable as examples of how to apply these metrics in other areas. For example, calculating the food sustainability metric for an area can tell how closely local and sustainably produced food can come to meeting food demands. However, it also quantifies how much additional food, as well as the relative amounts of different food types, that are needed to meet a population's food needs. This is very important information to know when one wishes to encourage production of certain types of agricultural products to meet residents' needs and to achieve a more efficient local food system.

The above CBAs, combined with the food sustainability metrics, are powerful analysis tools to examine different development and production scenarios, as well as guide policy and planning efforts. For example if a local government wishes to invest money into a sustainable agriculture initiative, it may wish to use a CBA to calculate the estimated startup and operations costs of a single operation. Using this information, the number of community gardens that may be founded with $100,000 of grant money can be calculated. If appropriate data are available, the local government may be able to estimate the net community benefit that could be realized by this investment (i.e. the return on the investment) in terms of food produced and other effects such as energy conservation and economic stimulus. If these data are not available, they could be collected over a several year period by the funded gardens and used to calculate the benefit they provided. Once the amount of food that could be grown or is being grown in known, the sustainability metrics can be used to understand how these funded gardens are increasing community food sustainability in a tangible way. These metrics can also inform how much more would be achieved with additional investment. For example, if 1 percent of the residents of an area grew a food garden of 100 m², what would the expected productivity be and how would that increase food sustainability for the community?

Household Food Sustainability Analysis

An analysis of food production data obtained during the residential food growing operation study was conducted to determine how much the operation contributed to household food sustainability. This was conducted by determining how closely the harvested food came to providing for a healthy diet for the household. The sustainability metric used in this analysis used the Choose My Plate guidelines (Table 17) along with the assumption that *if sufficient food could be grown on the residential home's property to meet the Choose My Plate guidelines for a healthy diet, then the household would achieve a high level of food sustainability, including food security, food sovereignty and food resilience.* It is recognized that this exercise omits other important food groups that are needed for a healthy diet. However, these methods can be applied to those food groups once adequate data can be obtained for the analysis.

Conversion factors between the harvested weight of crop types and the number of servings (cups) of different classes of vegetables and fruit (Table 17) were used to calculate the total food servings that were produced by this backyard garden (Table 44). These results indicate that that the residential food garden could produce enough dark green, red/orange, other vegetables and fruit to meet the annual dietary needs of two adults. The food garden produced almost enough dark green vegetables to meet annual dietary recommendations for six adult males and relatively small amounts of starchy vegetables and beans/peas.

Some of the challenges of growing starchy vegetables during the study included poor potato production in this climate. Tropical vegetables were not part of the food heritage of the household, which also lacked knowledge related to growing regionally-appropriate starchy vegetable crops; both of these factors were barriers to growing climatically-appropriate crops. One tropical starch vegetable that is widely grown in the region is cassava, which would have made a better starch crop for the residential garden.

Table 44. Number of servings of fruit and 5 categories of vegetables harvested during the residential food garden study. Reported values exclude coconuts, which were harvested for water.

	Vegetables					Fruit
	Dark Green Vegetables	Red & Orange Vegetables	Starchy Vegetables	Beans and Peas	Other Vegetables	
2009-10	157	119	62	22	474	12
2010-11	331	366	98	46	734	470
2011-12	518	534	95	30	782	1076
2012-13	480	626	77	92	536	1617
2013-14	828	686	38	46	712	152
2015	n/a	n/a	n/a	n/a	n/a	1580
3-year Mean	609	615	70	56	677	1424*
# Adult Male Diet Needs**	5.9	2.0	0.2	0.2	2.6	2.0

**The 3-year mean for fruit excluded the 2013-2014 extremely low fruit production year, which was consistent with the calculations in Table 26.

*The number of middle-aged adult males whose dietary needs could be met by the food harvested from the residential food garden study; middle-aged adult males were chose as the conservative measure since that demographic requires the largest food intake, as described by the Choose My Plate guidelines.

According to Purdue University (2017), subsistence farms produce approximately 9.8 tons of cassava per hectare of land. Using this productivity rate, it was calculated that approximately 20 m² (215 ft²) of garden space would need to be dedicated to cassava production, an area that may be obtained by growing fewer dark green vegetables.

Beans and peas were not produced in high enough volumes over the study period to provide a major portion of what would be needed to meet annual dietary recommendations for an adult. As with tropical starchy vegetables, tropical bean and pea varieties were not part of the food heritage of the household, which also lacked knowledge related to growing these crops. Pigeon peas, black beans and black-eyes peas produce well in this region and a heavier focus on producing these crops would have significantly raised productivity of this vegetable category.

Although a large number of varieties of vegetables were grown and harvested during the study period, fruit production was dominated by the highly productive mango tree. From a practical perspective, enough fruit volume was grown to feed two adults; but no one would prefer to or should consume a diet that consisted of predominantly one type of fruit. Excess fruit beyond what is preferred by the household could be sold or traded for other fruits to increase food choices. Also, several fruit trees that were installed at the onset or during this study had not yet matured. It is anticipated that within several years, the ratio of mangoes to other fruit produced will become more balanced.

Results from this study suggest that by adjusting the amount of space dedicated to each category of vegetables, by planting crops that are more suited to the region and planting more varieties of fruit, the household may well be able to produce all of the servings of fruits and vegetables needed by the its two adult residents. Further studies would be needed to determine specific crop types, planting schedules and amount of space that should dedicated to each crop to grow the produce needed to feed the household.

Although it may be possible to meet the vegetable and fruit food demand for the household from this residential lot, this is not enough to meet the full food demand for the household. Other food types necessary include grains, protein foods, dairy and oils. These were not grown during this study and it is not known how much of these could be produced within the residential garden footprint. Other studies would be needed to address these food groups, as they lie beyond the scope of this study.

Community Food Production and Food Sustainability

Food production data from the residential food growing study was used to calculate (through extrapolation) the volume of food that could potentially be grown within the metropolitan area's residential properties. Using food production data, the area of available growing space within the metropolitan area's residential development and census data, this analysis sought to calculate how closely available food growing space may be able to meet local food demand for vegetables and fruit. It was assumed that if sufficient food could be grown within the metropolitan area's residential properties to meet the Choose My Plate guidelines for a healthy diet, then the metropolitan area would achieve a high level of food sustainability, including food security, food sovereignty and food resilience. It would also provide a practical means to eliminate local food deserts.

A random sampling of quarter-Sections within the West Palm Beach metropolitan area was examined to determine the amount of potential food growing space. The total extent of sampled area was 0.94 percent of the entire metropolitan area, which contained a variety of land cover types. The metropolitan area consisted of 463 sections (each being approximately 1 square mile in extent) and the total extent of the metropolitan area was approximately 119,917 hectares.

Within the sampled area, 1916 single-family home properties and 431 multi-family unit properties (a total of 2347 residential properties) were examined. The amount of growing space available within the sampled residential areas was 575,237 sq. meters (57.5 hectares) or approximately 5.1 percent of the sampled area. Because the randomly-sampled residential area was assumed to be representative of residential development patterns in the entire metropolitan area, the total extent of growing space available in the West Palm Beach metropolitan area was estimated to be 61,157,635 sq. meters or 6116 hectares.

Potential food production (FP_p) on the estimated available residential growing space was calculated from the food production data obtained from this study (3-year averages from Table 44) and the area of the food growing operation (food production area + utility area). The FP_p for metropolitan West Palm Beach is shown in Table 45. These large numbers demonstrate how much food must be produced each year to feed a population of approximately 1.3 million people. When one considers the entire population of a country, the food needed to feed so many residents is staggering.

Table 45. FP_P for metropolitan West Palm Beach, Florida, United States.

Produce Type	PFp (Number of Servings)
Dark green vegetables	438,176,467
Red & Orange vegetables	442,493,477
Starchy vegetables	50,365,111
Beans & Peas vegetables	40,292,089
Other vegetables	487,102,575
Total vegetables	1,458,429,719
Fruit	1,024,570,262

The potential food demand created by the metropolitan area's residents (the FD_p, for fruit and vegetables only) was calculated from the population's age and sex demographics and the amount of produce (fruit and vegetables) that would need to be consumed by them to meet the Choose My Plate guidelines for a healthy diet. According to the 2012 U.S. Census data, the metropolitan area's population was 1,328,553 people of which 48.4 percent were female. Census data from were used (instead of more recent census data) because it more closely corresponded to the aerial photo analysis timeframe, which was used to calculate the potential food production values. The FD_p values for metropolitan West Palm Beach are shown in Table 46.

The Food Sustainability Measure (FS_M) values for different vegetable and fruit types that would result from producing food in residential backyards in the same manner as it was produced in the residential food garden study are shown in Table 47. Under this scenario, there would be a production deficit of over 300,000,000 servings of starchy vegetables and almost 75,000,000 servings of beans and peas per year. In contrast, the amount of Dark Green vegetables and Other vegetables significantly exceeded what the population within the metropolitan area needed.

Table 46. Palm Beach County Metropolitan Area's FD$_P$; values are weekly number of servings.

Age & Sex	Vegetables					Fruit
	Dark Green	Red & Orange	Starchy	Beans & Peas	Other	
<5 years F & M	55,226	202,494	202,494	36,817	147,269	515,440
5-18 years F M	160,113 174,514	507,026 573,404	480,340 548,474	133,428 174,514	400,2284 448,751	1,120,794
18-65 years F M	644,933 803,359	2,364,753 2,410,077	2,149,775 2,410,077	644,933 803,359	1,719,820 2,008,397	6,019,371 5,623,513
>65 years F M	257,525 240,589	686,734 882,161	686,734 801,964	171,683 240,589	600,892 641,571	1,802,676 2,245,500
Weekly total	2,336,260	7,626,649	7,279,859	2,205,324	5,966,985	18,548,895
Yearly total	121,485,501	396,585,754	378,552,665	114,676,836	310,283,211	964,542,553

The FS$_I$ values, which are an expression of the percent of the total demand that is being met, indicate that the food production in the metropolitan area could be met for several types of vegetables and for fruit. By changing the amount of area dedicated to different vegetable types, it would be possible to produce greater quantities of Starchy vegetables and Beans & Peas to better meet food demands. It is important to recall the discussion above about why Starchy vegetables were under-produced and that changing to crop types that are more appropriate to the region would need to be a part of the solution. In addition, this analysis only examined food-growing space within residential neighborhoods and is a conservative estimate. If the analysis included other potential food growing areas within the metropolitan area, such as institutions, community gardens, parks and commercial land, the productivity estimates below would surely increase and suggest that it may be possible to produce enough, or nearly enough, of these food types to meet the guidelines for a healthy diet for all of the residents within the metropolitan area.

Table 47. FS_M and FS_I values for different food types that would results from growing produce in existing growing space in residential backyards within the West Palm Beach metropolitan area.

Produce Type	FS_M (Number of Servings)	FS_I
Dark green vegetables	316,690,966	3.6
Red & Orange vegetables	45,907,723	1.1
Starchy vegetables	-328,187,554	0.1
Beans & Peas vegetables	-74,384,747	0.4
Other vegetables	176,819,364	1.6
Fruit	60,027,708	1.1

Results from this analysis show how food production on urban greenspace could benefit a community. The ability to quantify potential fresh fruit and vegetable production is important for understanding how much demand can be met locally. It also helps to visualize how different types of urban agriculture (private gardens, community gardens or commercial agriculture) may contribute to meeting food needs in cities.

Urban greenspace can produce significant quantities of fresh vegetables. For example, Ghosh (2010) reviewed the results of the Australian Bureau of Statistic's backyard gardens survey; which estimated that an average backyard grew 70.4 kg (155 lbs.) of vegetables per year (Australian Bureau of Statistics 1992). Head and Muir (2007) also studied Australian backyard gardening and found that some 52% of urban residents had backyard gardens.

The cumulative areas of suburban gardens and greenspace constitute the largest single urban land-use type (Randall *et al.* 2003, Gaston *et al.* 2005). Backyard gardens may represent the greatest potential area for local food production in most metropolitan areas, because of the development pressure on urban, suburban and peri-urban farms (Aubrey *et al.* 2012) as well as the limited availability of urban land for agriculture. As the number of people living in cities continues to increase, dependency on large-scale and distant agriculture will only grow.

Although the dominant urban food flow models rely heavily on food imports from outside of its region, this is not a necessary requirement to meet a city's food needs. For example, England's Victory Gardens met half of the nation's fruit and vegetable needs during and after World War II (Garnett 1996, Martin & Marsden 1999). During this time, most major cities in the United States (e.g., Boston, New York, Chicago) established Victory Gardens on public lands and the number of private backyard vegetable gardens reached approximately 34 million (Alexander *et al.* 1999). More recently, nearly all of Havana's (Cuba) vegetables and most of its fruits come from within a 30-mile radius of the city (Altieri 1999).

Chapter 15. Towards a Verifiably Sustainable Agricultural System

The sustainability science is a relative newcomer to the quantitative sciences. It will take the efforts of many individuals in the disciplines of research, philosophy and planning to grow and mature the field over the coming decades. The involvement of these very different disciplines is not contradictory; in fact, it is necessary. There can be no subject area that overlaps the hard sciences, the humanities and the applied sciences more than food. Through the process of maturing, new things will come along to replace other things that have long been held dear. That will make it an emotional and sometimes difficult journey.

Today, the sustainability literature is still very much dominated by descriptive studies. That is changing. The shift towards a more quantitative approach will ultimately yield a more verifiably sustainable agricultural system. In order to fully achieve this goal and to facilitate its establishment, it is necessary to look beyond the farm field and the enumeration of environmental effects. Agricultural systems are much more complex than that. It is generally understood how to determine fertilizer needs, crop planting schedules, irrigation needs and best harvest practices for a food growing operation that will achieve a high level of sustainable resource use. But what constitutes sustainable use differs from region to region, so sustainability guidelines must be revised and adapted to specific areas. Even still, if a proposed food growing operation were to adapt environmentally sustainable practices, develop a business model to assure economic sustainability and seek to fulfill social needs within the community, the agricultural operation could still fail before it planted its first crop because of external factors that lie well beyond the practice of growing food. These factors include cultural preferences, policy and politics.

When Sustainable Agriculture Science meets Culture, Policy and Politics

Scientific research is based on the practice of questioning beliefs and uncovering truths using empirical data, with the goal of using that information for a purpose. The scientific method has been founded on the concept that knowledge evolves and ideas that are demonstrated to be false are discarded in favor of those that have been validated. This approach is often at odds with culture, policy and politics, which have other agendas and apply different standards when evaluating the veracity of beliefs.

These differences cannot be ignored. In fact, it will never be possible to obtain a sustainable agricultural system unless bridges are built between them. More than one sustainable urban agriculture project has died when its ship crashed into the cruel reef of cultural, policy or political resistance. More details on these are provided below.

Culture

Food plays an enormous role in peoples' lives and is a large part of many cultural activities. Even though certain foods may be unhealthy to consume in more than modest quantities over a long period of time, this does not deter many people from consuming them in amounts that adversely affect their health. This fact is often overlooked when well-meaning organizations seek to improve food access in food deserts, or to increase food security in low-income communities, or when local organizations establish community gardens that fail after only a few seasons because of a lack of interest on behalf of local residents. Even if an extensive library of robust agricultural studies existed that pointed a clear path towards food sustainability, that alone will do little to increase the practice of sustainable agriculture. One reason for this is that the influence of culture is strong. Cultural practices and beliefs are engrained in a person's behavior and are perceived to be time-tested. For most people, these practices and beliefs work, so there is no drive to change them. Regardless of whether a scientific study demonstrates that one practice is superior over another, if there is cultural resistance to such a change, the new practice will be difficult (if not impossible) to implement. This difference between scientific fact and cultural knowledge is what creates the gulf between what people should eat (for a healthy diet) and what they do eat.

Although the residential food growing study presented in the previous chapters found that it was feasible, with adjustments to planting schedules and crop types, for the household to become highly food sustainable (with respect to fruits and vegetables only), this cannot be achieved unless the household is willing to adopt vegetable varieties that can be grown in the regional climate. In fact, the main barrier to growing all of the vegetables needed to supply the household during the study period was the desire to grow varieties that were culturally familiar and desired, but not necessarily climatically recommended. For those who are unwilling to change their food preferences to include varieties that can be produced locally, there is *cultural resistance* to food sustainability.

The idea of cultural resistance to food sustainability is important. Farmers and non-commercial growers who employ fully sustainable food growing methods but produce varieties that are not aligned with the local food culture will be unable to sell their crops to local residents. This problem underscores the necessity for sustainable agricultural operations to understand the cultural food needs of local residents. However, it will also be necessary for local residents to incorporate food varieties that can be reasonably produced by local operations. This is not unachievable. Commercial agriculture is constantly introducing new varieties of fruits and vegetables through grocery stores. Some of these are eventually incorporated into local food cultures and provide a means for invigorating an otherwise commonplace menu.

Another aspect of cultural resistance to food sustainability occurs when a certain cultural group has no affinity for or relationship with agriculture. For them, the food appears at the grocery store and they do not particularly care where it came from, how it was produced or why that would be important. These residents do not wish to participate in food growing and are content to pay someone else to do that for them. With this loss of connection to food growing, residents have less knowledge about food origin, how food is produced and food safety with regards to pesticide and fertilizer use (Rees 1992, Pothukuchi 2004). Interviews with study participants who felt this way can be found in Zahina-Ramos (2015). It is important to understand their perspective because sustainable urban agriculture will be most successful when the community is engaged with and interested in supporting local food production.

During the community survey that was conducted in tandem with the residential food growing study (Zahina-Ramos 2013), the question of how residents felt about growing food was asked. Feelings about backyard food gardening among non-food growers were primarily positive. Less than 10 percent of non-food growing respondents indicated that they did not like gardening or felt that they would rather do other things (Figure 26). More than 55 percent said that they felt it was a good idea and more than 40 percent said that they had considered trying it. This finding indicates that, in the area surveyed, receptivity to backyard food growing among non-food growers is relatively high and that only a small percentage did not like food gardening. Results from this survey indicated that there was little resistance to sustainable urban agriculture, particularly at the personal level. But this is not always the case at the organizational or governmental level.

Policy

Policy is used to achieve goals and to guide activities towards a desired outcome. Policy is especially important for guiding community development patterns and forms, as well as organizing human settlements. Community food production is well within the realm of urban policy, as policy can play an essential role in local planning for food systems, promoting linkages between food production and community objectives, building community relationships and giving citizens a feeling of empowerment and control over meeting their own needs. Close proximity of food production sites to living space makes agriculture visible to city residents on a daily basis. This can lead to a greater concern for maintaining a cleaner urban environment, more sensitivity to the realities of food production, connection with the land and other positive effects.

Feelings about Backyard Food Gardening

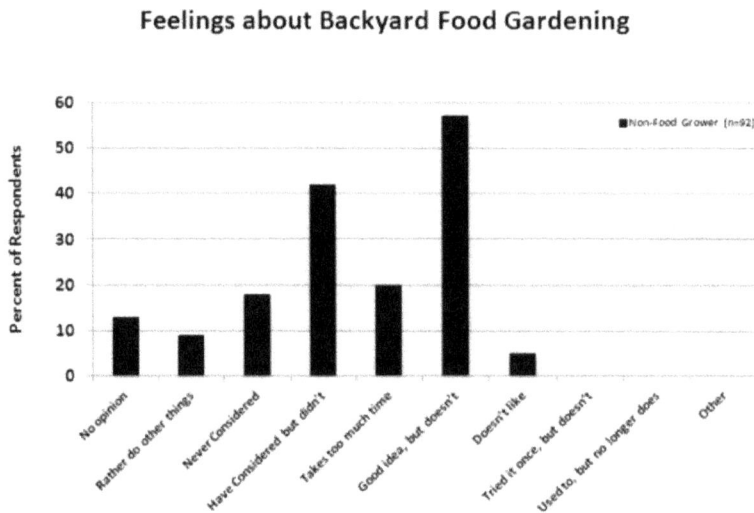

Figure 26. Questionnaire respondents' feelings about residential food growing. Figure source: Zahina-Ramos 2013.

Urban agriculture can be highly influenced by local-scale and regional-scale governmental policy. Municipalities may restrict or welcome urban agricultural operations, based on community and institutional standards. Even though sustainable urban agricultural operations may be desirable, without policy support they will not be allowed to operate (or will be difficult to operate) within the jurisdiction of a governing body. This roadblock to sustainable urban food production is called *policy resistance* to food sustainability. Policy barriers may be changed, but it takes perseverance and rallying local support for these changes.

Examples of policies that are barriers to sustainable urban food production include:

- Homeowner association or neighborhood association bylaws that restrict landscape content or what activities can take place within their jurisdiction.
- Governmental zoning, rules, ordinances or laws that prohibit food production in specific areas.

- Nuisance laws that indirectly restrict or ban certain activities that are related to food production. Some of these laws are based on noise, aesthetics or odor.
- Exclusive rights policies that restrict access or activities to select entities.
- Selective or targeted programs that support unsustainable agricultural practices over sustainable agricultural practices; some examples include agricultural subsidies.

Besides these formalized policies that can be barriers to urban agriculture, there are other obstacles that are unintended results from policy decisions related to zoning and development patterns. Recent trends in urban design have focused on development that utilizes fewer resources, less space and is more efficient (Berke *et al.* 2003). This type of development is patterned after pre-20th Century cities that were built on a human scale. One example, New Urbanism, promotes: 1) higher-density urban development and infill of existing urban land to reduce the per-capita energy and material resources consumption, 2) reduction of the amount of paved area, 3) increased ratio of open space to housing units, 4) reduced driving distances and 5) making mass transit more efficient (Rees & Wackernagel 1996, Berke *et al.* 2003).

Smart Growth, as described by Downs (2005), uses the following principles: 1) make existing settlement more compact by limiting the outward extension of new development, 2) increase residential densities in new and existing neighborhoods, 3) allow for more mixed land use types and pedestrian access to minimize the use of personal vehicles for short trips, 4) support the public costs of new development through impact fees, 5) promote the use of public transit, and 6) revitalize older neighborhoods.

However, there is not broad agreement that New Urbanism and Smart Growth models are desirable. Gordon & Richardson (1997) question whether compact cities are a desirable planning goal, since low-density settlement is the overwhelming choice for residential living. Other problems include:

- High-density development within cities further distance residents from agriculture and knowledge about how their food was grown.
- Promoting infill of vacant urban land eliminates potential food growing sites within cities.
- These development models exclude food production areas, leaving the issue of sustainable urban food production unaddressed.

Because of existing municipal regulations in the United States, urban agriculture is mostly practiced as fruit and vegetable gardens and rarely includes fowl, livestock or fish (Brown & Jameton, 2000). However, some municipalities have moved to better facilitate the practice of more types of agriculture within their jurisdictions. The cities of Baltimore, Madison, Milwaukee, Seattle and Washington, D.C. have significant community garden initiatives and allow poultry raising within city limits (Broadway 2009). Although initiatives and policies are mostly at the city level, a national urban agricultural system in the United States does not exist. According to Peters (2010), a move toward a national urban agricultural policy would require the acquisition, dedication and protection of urban land for community gardens along with appropriate zoning ordinances.

The establishment and support for sustainable urban agriculture requires the following:

- Access to land
- City residents value local food production and see it as an essential service
- Support from local citizens
- Leadership from local authorities
- The development of an integrated planning policy that considers the economic, environmental and social benefits of sustainable urban agriculture
- Access to local food supply distribution systems

Politics

As everyone knows, things can change significantly with one governmental election. When a change candidate takes office, sweeping alterations are often made to policy direction, programs and laws. These can either adversely affect or benefit sustainable agricultural operations.

Laws are usually enacted to benefit a certain program or sector of society. Agriculture is often viewed as being an essential service that should be supported for several reasons; some of these being: the importance of maintaining an affordable food supply, the need for reliable food production, support for increasing efficiency of operations and the importance of the food system to the economy. Many proposed laws are subjected to a period of consideration, discussion and debate before being passed. During this period, lobbyists and the public are invited to voice support for or concerns with a particular proposed law. Large agribusinesses are better positioned to represent and support laws that favor them than small-scale or non-commercial food growing operations. Because of this, sustainable agriculture does not benefit from the subsidies, price supports and other incentives that many large-scale agribusinesses do. This creates a system that promotes economic viability of unsustainable agricultural businesses over that of sustainable agricultural operations. When politics work against the ability to establish and expand the practice of sustainable agriculture, this is referred to as *political resistance* to food sustainability.

Political support for urban and sustainable agriculture is essential for sustainability because it supports resilience. Resilience is the aspect of a sustainable agricultural operation or food system that allows it to continue to provide in the face of disruptions. Disruptions are often viewed as being climatic in nature, but disruptions can also be caused by political uncertainties and policy changes can be difficult to weather. Political commitment to sustainable agricultural systems is essential if resilience is to be realized. This commitment can only come when sustainable urban food growers organize and give voice to their concerns, and elevate their needs to politicians. Involvement in the political system is necessary for a sustainable agricultural system to remain economically and socially viable…it is also necessary for a healthy democracy.

The Pendulum and Patience

Pendulums swing. It's why they were constructed. Initially invented to run a timepiece, they are now used for other applications such as detecting the movement of the earth's tectonic plates. As the bob moves towards an extreme, the natural force of gravity decelerates the tethered weight and draws it back to the center. With each swing cycle, another second in time advances. Pendulums were replaced by other, more efficient, technologies almost a century ago- around the time that industrialized agriculture became widespread and replaced the ancient agricultural traditions that once dominated the food growing methods practiced in developed countries.

Many social movements, over the long term, behave like a pendulum, swaying from one side to another but ultimately returning to a more centrist position. Movement towards one extreme is driven by the social drivers of the moment. These drivers are fear, anxiety, the desire to do right and the hunger for justice. What keeps the bob of social movements tethered to the pendulum are the local and national discussions that take place, which draw in all sides of the issue for consideration and debate, and reposition the bob back towards a balanced center.

With respect to food production in developed countries, over the 20th century the pendulum bob was driven towards an extreme of industrialized food production and away from local-scale sustainable agriculture. This drive was fueled by a fear of food shortages and hunger caused by growing populations, by the anxiety left behind by the Great Depression that saw many people malnourished or die from a lack of food, the desire for a productive and abundant food system, and for equitable access to affordable food. In so many ways, the industrialized agricultural system of the 20th century achieved that.

Then, the pendulum bob momentarily paused at the extreme where industrialized and unsustainable food growing practices were seen as the pinnacle of human achievement. Then, it began to descend towards the center again, forced there by the discovery of previously unknown and unintended consequences resulting from unsustainable agricultural practices. The bob is now moving back towards the sustainable agricultural practices that were abandoned more than a century ago. This time, the bob is driven by the fear of destroying the resources humans must have in the future, anxiety about the impacts of pollution and environmental degradation resulting from unsustainable agricultural practices, the desire to do right and the hunger for a food system that does not exclude those who cannot pay to participate.

It is not enough to know that public awareness and opinions related to agricultural sustainability change. This change only happens because of active participation by scientists who collect data and report findings, by citizens who voice concerns and by community leaders who are responsive to issues. This process also requires diligence and patience. Lots of patience. Humanity is facing a highly uncertain and precarious future with respect to its food supply, if changes are not made. It is the nature of unsustainable systems to become obsolete. Because of this, sustainability is not an option nor is it a luxury, it is a necessity.

Glossary and Abbreviations

ACGA. American Community Garden Association

Agricultural Census. A census of agricultural crops, production and acreage in the United States conducted by the United States Department of Agriculture on years that end with a "5".

Agriculture. The practice of cultivating organisms for products, such as food, fiber, fuel and medicines.

Agrochemicals. Chemicals used in the cultivation and harvest of food crops. These can include synthetic fertilizers, pesticides, hormones and growth agents.

Agroecology. A type of agriculture that integrates natural ecological processes and principles in the crop production activities. Agroecological practices may include no-till farming and organic agriculture.

Agronomy. The practice of growing and using plants for various purposes, such as food, fuel, fiber, dyes or other materials.

Analyst. The person or entity responsible for conducting the CBA.

Aquaculture. The practice of growing food crops within an aquatic environment, e.g., fish.

Aquifer. The pool of underground water that is stored in geological formations and may be extracted for use. Ground water, in some cases, may be recharged by rainfall or seepage from rivers or lakes.

Aquifer Depletion. The unsustainable condition where aquifer withdrawals exceed aquifer recharge.

Benefits. Products, crops or positive effects resulting from conducting an activity.

Best Management Practices. Management changes that improve the sustainability status , or reduce environmental impacts, of a particular practice.

Biodiversity. The sum of all life within a defined space. Biodiversity can be measured (or described) at the genetic, species or ecosystem scale.

BMP *See* **Best Management Practices**

Carbon Dioxide. A type of gas (CO_2) emitted from combustion processes and animal respiration. One gallon of diesel fuel produces approximately 10.2 kg (22.4 lbs.) and one gallon of gasoline produces approximately 8.9 kg (19.6 lbs) of CO_2 emissions when burned.

Case Study. A research method that involves intensively or comprehensively studied factors acting on a single or few subjects, as well as the context which with the subject is (or subjects are) placed. The value of case studies is in the detailed examination of mechanisms, functions and relative contribution of factors, rather than a characterization of the population.

Cash Receipts Sales. The amount of money that the commercial farmer receives through the sale of harvest.

CBA. *See* **Cost-Benefit Analysis**

CEA. *See* **Cost-Effectiveness Analysis**

Choose My Plate. A set of dietary guidelines defined and published by the United States Department of Agriculture.

Civic Agriculture. Local-based, small-scale agriculture that involves or encourages citizen interaction.

Climate. The average weather conditions over a 30-year (or longer) period of time.

Climate Gradient. Changes in temperature, moisture or other climate parameters that occur over a distance and can be caused by changes in environmental conditions.

Community Garden. A type of civic agriculture where a single piece of land is collectively gardened by many people, usually residents of the local community. Land may be publicly or privately owned, and garden plots may be individual or shared.

Complex Landscapes. Areas that have a high degree of heterogeneity with respect to environmental conditions. Environmental conditions in the urban landscape are modified by: (1) the physical urban form, which arises from development patters; (2) ecological gradients and patterns of greenspace; and (3) environmental gradients and resources.

Container Agriculture. A type of agriculture where crops are produced in containers, rather than in the ground (i.e., a field or garden bed).

Conventional Agriculture. This term usually refers to crop production using industrialized methods (see Industrial Agriculture).

Cost-Benefit Analysis (CBA). A type of analysis that attempts to identify, quantify and estimate the relative costs and benefits of an activity. It can also be used to compare between alternative operating proposals or to examine with- or without-project effects. Monetary values are time-adjusted in the CBA.

Cost-Effectiveness Analysis (CEA). An alternative to the **CBA** where values may be placed in physical units instead of monetary units.

Costs. Impacts, investments, inputs or negative effects involved with conducting an activity. The sum of these for an operation is the Gross Costs.

Crop Futures. Futures contracts are agreements between parties that consent to buy and sell an agricultural crop for a pre-determined price, usually before the crop is harvested.

Crop Rotation. The practice of managing crop selection, placement and planting schedules to avoid agricultural pests and increase soil quality.

CSA. Community Supported Agriculture

Desertification. The process of relatively dry land becoming increasingly arid and causing ecological change toward xeric or desert conditions.

Diesel Fuel. A type of liquid fuel that is make from crude oil (petroleum), other fossil fuels (such as natural gas or coal), biomass or vegetable oils. Diesel fuel is used to run some types of machinery and vehicles. Approximately 12 gallons of diesel fuel are produced from one 42-gallon barrel of petroleum. One gallon of diesel fuel produces approximately 10.2 kg (22.4 lbs.) of CO_2 emissions when burned.

Direct Benefit. Benefits that are directly realized from an activity; e.g., a crop.

Direct Cost. Costs or effects that are explicitly input to the operation. These include labor, materials, water, fertilizers and monetary investment.

Discount Rate. Changes in the value of currency over time results in inequalities between the value of $1 between years. Because of this, a discount rate is applied to bring all past currency values into the present currency value.

Ecological Farming. A type of agriculture that seeks to preserve and restore many of the ecological goods, services and functions that are found in natural systems while producing an adequate and sustainably grown suite of crops.

Economics. The study of the production, distribution and consumption of goods (tangible commodities) and services.

Educational Garden. A garden that has been established for educational purposes, often located on the grounds of an educational facility.

Effects. Changes caused by or resulting from an activity.

EIL. Economic Injury Level, the minimum level of harm produced by an agriculture pest that would require active management. Below this level of harm, there is no reduction in crop productivity.

Environmental Costs. These are impacts to the environment resulting from conducting an activity.

Exclusion Bias. When conducting an analysis, results can be skewed (biased) when certain factors have been excluded from consideration.

Expenses. Monetary investment involved with conducting an activity.

Extended Costs. *See* **Indirect Costs**

External Factor. A factor that lies outside of the operation or activity (and is beyond its control), but may by influenced by it. An example is governmental agricultural policy.

Extreme Heterogeneity. An environmental condition that occurs in cities where different microclimates are formed by variations of wind exposure, lighting, moisture and temperature conditions from site to site.

Farm. An area of land that is used to conduct agriculture.

Fixed Costs. Costs that must be incurred by the farmer whether or not production takes place.

Food. The portion of a crop organism that is consumed for nutrition. The parts of the crop that are utilized are often culturally defined.

Food Demand (FD). The amount of food needed to provide for a healthy diet, expressed for different food types. The potential food demand for a defined population (FD_p) is the amount of food needed (expressed as different food types) that would provide a healthy diet for that population.

Food Desert. A geographic area where residents have limited or insufficient access to healthy, fresh and nutritious food.

Food Insecurity. A condition where an individual or population has insufficient access to nutritious food, either because of a lack of supply or inability to acquire it.

Food Production (PF). The amount of food produced in an area or foodshed. Potential food production (PFp) is the amount of food that could potentially be produced from an area or foodshed, based on estimates. The historic food production (PFh) is the actual amount of food produced from an area or foodshed.

Food Sustainability. This is achieved when sustainably produced food is available in the quantity and types needed to provide for a healthy diet for all of the residents of an area. Food sustainability can be conceptually defined as the ability to meet the dietary needs of a population through locally and sustainably produced food.

Food Sustainability Measure (FS_M). The difference between the amount of food that is or can be (locally and sustainably) produced and the amount of food needed to provide for a healthy diet for a population.

Food Sustainability Index (FS_I). The ratio between the amount of food that is or can be (locally and sustainably) produced and the amount of food needed to provide for a healthy diet for a population. When the FS_I is equal to or greater than 1, the population has achieved food sustainability.

Food System. The entire chain of events, processes, steps and parts involved in the production, transport, marketing, consumption, disposal and social aspects of food.

Foodshed. The geographic area where food moves from its point of origin to its market or consumer. It is the area that contains all of the activities that produce food for a target population.

Fossil Fuel. Fuels derived from ancient geological formations that are believed to have originated from organic sources. These are high in hydrocarbons that when burned, give off CO_2 emissions and other pollutants. Examples of fossil fuels include coal, petroleum and natural gas.

Gas (Gasoline). A type of liquid fuel used to power some types of engines and machinery. Approximately 19 gallons of gasoline can be produced from a 42-gallon barrel of petroleum. One gallon of gasoline produces approximately 8.9 kg (19.6 lbs) of CO_2 emissions when burned.

Greenhouse Gas. One of any gasses that, when present in measurable concentrations in the atmosphere, retain heat that would otherwise be lost from the system. Examples of greenhouse gasses include CO_2, water vapor, methane and nitrous oxide.

Greenspace. The space in urban areas that are not covered by pavement or development. In urban areas, this space is typically lawn, parks or open space.

Gross Benefits/Costs. The total sum of all benefits or all costs from conducting an activity.

Growing Season Duration. The average length of time during a year in which crops can be grown. The definition of growing season varies by country and is influenced by latitude and elevation.

Guerilla Garden. A garden that has been established on land that the gardeners have no legal right to use.

Hidden Drought. Drought conditions that can occur outside of an extended dry period, resulting from a series of light rain events that do not provide moisture to plants or high evapotranspiration (combined evaporation and transpiration) rates.

Historic Food Production (PFh). *See* **Food Production**

Hydroponic Agriculture. A type of **industrial agriculture** where crops are produced using a soil-less medium.

Indirect Benefits. Benefits that are indirectly realized from an activity, such as a secondary effect.

Indirect Costs. Costs or effects that extend beyond the realm of the farmer and farm field. Some indirect costs include impacts from pollution emanating from the growing site, environmental degradation caused by the food growing practices, ecological loss as production land is taken from natural habitats, and the loss of ecological goods and services associated with the loss of natural habitat.

Industrial Agriculture. A type of agricultural practice that relies on intensive crop densities, crop monoculture, the use of synthetic pesticides and fertilizers, and the use of mechanized machinery for planting, maintenance and harvest.

Inputs. All expenses, investments or efforts involved with conducting an activity.

Intangible Benefits. Products, crops or positive effects (that are not physical or monetary) that result from conducting an activity.

Intangible Costs. Impacts, investments, inputs or negative effects (that are not physical or monetary) that result from conducting an activity.

Internal Factor. A factor that lies within an operation or activity. An example is the crop plant's demand for fertilizer.

Law of Diminishing Returns. A farming principle where the amount of incremental benefit decreases with additional inputs. For example, the amount of additional crop produced by the addition of 10 grams of fertilizer will be decrease with each additional increase of 10 grams.

Macroeconomics. This branch of economics involves the study of influencing policies, resources (e.g., capital, labor force), inflation and economic growth/decline on an entire economy.

Markets. Any of the ways in which economic goods or services are exchanged between other parties.

Methane. A gas (CH_4) that is sometimes a byproduct of agricultural operations that is approximately 30 times more potent as a greenhouse gas than CO_2.

Microclimate. A locally-modified atmospheric condition that is different from the surrounding area.

Microeconomics. The study of how individual elements interact and produce outcomes within a market.

Net Present Value (NPV). The value of past, present or future items or projected options in a CBA, expressed in the context of today's currency value.

NIMBY. **N**ot **I**n **M**y **B**ack **Y**ard; a type of activity that residents typically rely on but don't want to live next to.

Nitrous Oxide. A chemical compound (N_2O) that is 298 times more potent of a greenhouse gas than CO_2. It is emitted from fertilized agricultural soils, livestock manure and agricultural fertilizers.

Non-profit Farms. Food producing operations that are not part of the commercial market system.

Operating Costs. These are recurring costs that are required to continue an operation.

Opportunity Cost. The potential cost if a resource (e.g., money) is invested in one venture as opposed to being invested in another option. For example, a dollar invested in Option A may yield $1.03 but if invested in Option B, the dollar may yield $1.10. The Opportunity Cost for Option A is $0.07.

Organic Agriculture. Agricultural practices that rely on organically-derived fertilizers, natural pest control and crop rotation.

Outputs. All products, effects or results arising from conducting an activity.

Plant Productivity. A measurement of the amount (usually mass) of a certain plant product that is produced from a defined area over a specified period of time.

Potential Population Food Demand (FD$_P$). *See* **Food Demand**

Potential Food Production (FP$_P$). *See* **Food Production**

Proxy. *See* **Surrogate**

Referent Group. The person or entity who is relevant to the **CBA**; in effect, this is the customer who will use the analysis results to make a decision. The referent group orders the analysis and defines the parameters included in the CBA.

Salinization. A buildup of salt concentration in soil caused by the cycle of irrigation and evaporation; this is a particular problem in drier climates where irrigation water contains trace levels of salt. Salinization can cause soils to have levels of salt that are toxic to plants.

Scarce Resources. Finite resources with potential alternative uses. When using scarce resources, it is best that they are used for the most beneficial outcome.

Sensitivity Analysis. This analysis examines the amount of uncertainty (or variability) inherent in the results of a calculation or model output. This is achieved by changing input parameter values according to the range of potential (or expected) parameter variability and examining the net effect on output values.

Small Farm. A farm that has less than $20,000 annual sales (USDA 1978) or a retirement or residential farm that has less than $250,000 annual gross receipts (USDA 2010); the latter is the definition that is in common use today.

Small-Scale Farms. Farms that may be managed by small machinery, animals or by human labor.

Solarizing. The process of using clear plastic sheeting over agricultural soil during the hottest months to raise soil temperatures and kill soil pests.

Specialty Crops. Specialty crops are lower-demand crops that are grown in smaller quantities than traditional crops and may be targeted for niche markets.

Stakeholder. Entities who have an interest (e.g., monetary or legal) in or are affected by a particular activity.

Start-up Expenses. Monetary costs that are required to initiate an activity; these may include purchasing infrastructure and hiring expertise to create an operating plan.

STEM. Science, Technology, Engineering and Mathematics- a type of academic curriculum that includes these four areas of study.

Storm Water. Water that runs off of the land surface during a storm event, often carrying pollutants or dissolved materials along with it.

Subsistence Crops. Crops that are grown to provide for the complete food (and other) needs of the grower.

Subsistence Garden or Farm. A garden or farm that is grown to provide enough food (and/or materials) to feed (and/or clothe) the grower(s) throughout the year.

Surrogate. A data set, object or thing that can be used as a proxy (i.e., substitution) such as the value of a similar commodity or service.

Sustainability. A condition where resources are utilized in such a way that they are not damaged, depleted or exhausted.

Tangible Benefits. Products, crops or positive effects involved with conducting an activity that are physical or monetary.

Traditional Agriculture. A type of agriculture that has been practiced by indigenous peoples. This form of agriculture typically respects the carrying capacity of the land and is compatible with the local ecology.

Traditional Crops. Crops that are typically grown by commercial agricultural operations.

Unusual Crops. Crops that are grown for specific purposes, other than food. Some examples are worms for composting, bees for pollinating and ladybugs for aphid control.

Urban Agriculture. Agriculture that takes place within or adjacent to urbanized areas.

Urban Heat Island Effect. The phenomenon where diurnal temperatures within a city core are markedly higher than in the surrounding countryside. This effect is caused by the entrapment of heat by the urban infrastructure.

Valuation Theory. The theory that all things can be assigned some reasonably accurate value.

Variable Inputs. Inputs to an activity that change from time-to time. An example is rainfall.

Victory Gardens. World War II gardens that were promoted by the United States and Great Britain governments to produce food that required minimal investment of critical resources.

Willingness to Pay. A concept in **Valuation Theory** that something can be assigned a monetary value based on consumers' willingness to pay for that non-market item.

Index

A

Adaptive management · 74
Aesthetic value · 13, 57, 96, 101, 117, 131, 210, 227, 241
Agribusiness · 20, 25, 242
Agriculture (*also see* Commercial agriculture; Non-commercial agriculture)
  -Civic · 33, 77, 246
  -Conventional · 24-25, 29, 65, 108, 134, 187, 190, 202-204, 211, 219-221, 247
  -Large-scale · 22, 26, 29, 31, 34, 36, 54, 56, 79, 116, 129, 139, 190, 204, 235, 242
  -Microbial · 17
  -Small-scale · 29, 32, 54, 62, 64, 83, 129, 149, 158-160, 162, 190, 242, 246, 254
  -Subsistence · *see* Subsistence Farm
  -Sustainable · *see* Substainable Agriculture
  -Very small-scale · 62, 64
Agricultural
  -Census · 145, 147-148, 189, 245
  -Chemicals · 18, 68, 217, 245
  -Intensification · 9, 20
  -Pest predators · 35, 117, 196-197
  -Pests · 18, 23, 35-36, 43-44, 47, 54-56, 59, 71, 73-74, 76, 78, 82, 95, 195, 247, 254
  -Policies · 20, 37, 241-242
  -Productivity, *see* Productivity
  -Production data · 142, 145-146, 148-149, 151, 180, 230, 232-233
  -Products · 17, 19, 21, 26-27, 31, 41, 44-45, 67-68, 87-88, 91-92, 94, 138-139, 143-144, 146, 148, 154, 229, 245, 251
  -Type · 19, 22
Agrochemicals · 18, 68, 217, 245
Agroecology · 18, 22, 245
Agronomy · 17-18, 245
Air flow · 57
Amish · 23
Analyst · 83-84, 92, 122, 124, 126, 129, 131, 133, 143, 145, 150-151, 158, 229, 245
Animal agriculture · 17
Aquaculture · 26, 32, 245
Aquifer depletion · 12, 245

B

Backyard garden · 34, 94, 96-97, 169, 199, 230, 235
Backyard vegetable garden *see* Garden, Family/Home
Backyards · 28, 34, 178, 182, 233, 235
Balinese · 23
Beekeeping · 32
Beneficial organisms · 35, 41, 178, 196
Benefit-cost analysis (*also see* Cost- Benefit Analysis) · 121
Benefits (*also see:* Indirect benefits; Intangible benefits; Tangible benefits)
  -Of urban agriculture · 2, 9-11, 13-15, 19, 28, 32, 43, 69, 72, 87, 94-98, 100-101, 103, 105-106, 119, 121-123, 129, 132, 157-161, 163, 178, 193, 199, 202, 205, 207-208, 210, 213, 217, 221, 227
  -Measuring · 2-3, 10-11, 14, 41, 80, 92, 94, 99-100, 103-104, 121, 124-125, 128, 131, 133, 141, 150, 155, 157-159, 162, 209, 219, 223, 226-227
Best Management Practices · 13, 245
Biodiversity · 11-12, 23-24, 42-44, 59, 107, 116-117, 194, 199, 246
Brokers, agricultural products · 65, 68, 93

C

Carbon dioxide emissions · 80, 103, 109-110, 128, 135, 161, 172-173, 203, 217, 225, 246, 248, 250
Carbon footprint · 43
Carrying capacity · 6, 42, 255
Case study · 2-3, 81, 159-160, 162, 180, 186, 191, 193, 246
Cash outlay · 63-64
Cash receipts · 134, 149, 245
CBA, *see* Cost-Benefit Analysis
CEA, see Cost Effectiveness Analysis
Census data
  -Agricultural ·145, 147-148, 189, 245

-Population · 11-12, 145, 147, 152, 165, 167, 180, 183-184, 189, 232-233
Certified organic · 25
Chemical fertilizer , *see* Fertilizer, chemical
Chemical pesticides , *see* Pesticides, chemical
Choose My Plate · 91-92, 145-147, 152-153, 175-176, 230-233, 246
Civic agriculture · 33, 77, 246
Clean energy · 16
Climate · 12, 26, 28, 35, 41, 45, 47-50, 57, 68, 72-74, 77, 79, 82, 85, 100, 109-110, 118, 167-169, 172-173, 175, 186, 202-203, 209, 229-230, 238, 246, 249, 253
    -change · 12, 41, 100, 110, 203, 229
    -conditions · 46, 48-49, 53, 56, 71, 148, 185, 190
    -damages · 109, 110, 173, 175
    -gradients · 59, 246
Coal-fired electricity-generation, costs of · 109
Commercial
    -Agriculture (*including* Farming) · 9, 19, 20-22, 24, 29, 34, 36, 62, 64-68, 73, 81, 83, 87, 89, 91, 93-95, 132, 190, 204, 235, 239
    -Agricultural system · 20, 143, 157
    -Benefits of commercial agriculture · 20, 95
    -Economics · 9, 64
    -Farm (*including* Food-producing operations, *also see* Commercial rural agriculture *and* Commercial urban agriculture) · 8, 11, 13, 24, 29, 31, 54, 62-68, 71-72, 74-75, 81, 83-84, 87, 89-93, 104-105, 121, 132, 134-135, 138-139, 144, 147, 149, 160, 190, 204, 255
    -Farmer · 63, 73, 89, 93, 134, 246
    -Food supply/supply chain/supplier · 20, 32, 213
    -Food system · 20, 68, 143, 157
    -Large-scale agriculture · 2, 26, 29, 31, 34, 36, 54, 56, 79, 116, 129, 139, 190, 204, 235, 242
    -Markets/market system· 7, 20, 33, 39-41, 65, 68, 71, 74, 89-90, 93, 143, 252
    -Crops (*including* Plant products) · 25, 27-28, 88-90, 92-93, 148-149, 160
    -Rural agriculture · 13, 20, 41, 54, 67, 108, 159
    -Urban agriculture · 20, 61-62, 64, 67, 108, 159
Commercialism of food · 33
Community
    -Gardens/gardening (*also see* Garden, Community) · 11, 16, 19, 21, 25-26, 28-29,

31-34, 36-37, 43-44, 61-62, 64, 69, 75, 77, 81, 84, 87, 91, 93-94, 97, 129, 131-133, 138, 143, 149-150, 160-161, 221, 229, 234, 238, 241, 246,
    -Supported Agriculture (CSA) · 66, 97, 247
    -Survey · 161, 178-179, 205, 207, 211-212, 226, 239
    -Sustainability · 14, 35, 40, 42-44, 67, 69, 93, 95, 98, 101, 140, 154, 159, 209, 225-226, 230, 235-237
Complexity/complex landscapes · 1, 48, 56-59, 121, 247
Compost/composting · 23-24, 28, 30, 42-43, 63, 82, 84, 104, 115, 169, 171-172, 174-175, 177, 192, 201-202, 215-216, 218, 222-223, 255
Compostable materials · 15, 101, 113, 118, 163, 174, 201-202, 215, 225
Comprehensive efficiency CBA· 223-226
Comprehensive project CBA · 221-223, 225-226
Container food production · 22, 25-26, 29, 76, 79, 84-85, 133, 171, 191-192, 200-201, 216, 218, 247
Conventional agriculture · 24-25, 29, 65, 108, 134, 187, 190, 202-204, 211, 219-221, 247
Cost (*also see* Intangible cost)
    -External, associated with food production · 2, 9-11, 13, 19, 43-44, 62-63, 71-72, 80-81, 100-101, 105-116, 118-119, 127-128, 130-131, 138, 160-161, 163, 172, 175, 199-200, 202-205, 208-210, 213-214, 216-218, 220, 223-229, 247-249, 251-253
    -Internal, associated with food production · 19, 25, 29, 34, 61-66, 68-69, 71-86, 93, 105-107, 113, 127-128, 131, 159, 163, 177-178, 192, 199-200, 207, 209-212, 218, 220, 247-249, 251-252, 254
    -Measuring · 2-3, 9-10, 121-135, 138, 157-159, 162, 172-173, 209-210, 219-229, 246-247, 253
Cost-benefit analysis (CBA) · 121-134, 209-212, 219-229, 247, 252-253
    -Comprehensive efficiency CBA · 223-226
    -Comprehensive project CBA · 221-223, 225-226
    -Monetary-based CBA · 219-220
    -With-project CBA · 227-228
Cost effectiveness analysis · 131, 247
Cover crops · 24, 28, 56, 115, 118
Cradle-to-grave analysis · 10
Crop

-Diversification · 21, 24-27, 35, 41, 56, 75, 79, 117, 159
-Futures · 64-66, 92-93, 247
-Growing methods · 11, 21, 24-29, 31, 42, 45, 48-49, 53, 56, 65-66, 75-78, 159, 237, 247, 251, 253-255
-Plants · 21, 24, 27, 147, 167, 188, 231, 234
-Plant studies · 18, 83, 149, 151, 187, 231, 245
-Proxy/proxy data · 145-146, 149-150, 254
-Rotation · 18, 24, 78, 247
-Selection · 21, 27-29, 35, 41-42, 45-52, 54, 56, 68, 73-74, 76, 143, 164, 167, 169, 230-232, 234, 255
-Specialty · *see* Specialty Crop
-Value · 9, 20, 28, 41, 45, 52, 61-62, 64-65, 68-70, 72, 87-93, 117-118, 143, 159-160, 175-176, 187, 190, 194-195, 245, 248, 251, 255
-Yields · 9, 17, 20, 34, 42, 46, 52-54, 65-66, 78, 83, 88-91, 95, 113, 148-151, 185, 189, 210, 231-232, 234, 248, 252, 255
Crude oil · 172-173, 248
CSA · *see* Community Supported Agriculture
Culture · 4, 16-17, 22, 27-28, 35-36, 41, 48, 68-70, 88, 90-91, 95-98, 130, 137-138, 152, 159-160, 177, 194, 205, 207, 237-239, 249
Cultural resistance to urban agriculture · 238-239
Customer · 20, 65-66, 72, 92, 253

D

Day length · 49
Demand
-Agricultural products · 27-28, 41, 52, 62, 65, 89, 139, 143, 254
-Energy · 15, 214
-Food · 1, 6, 11, 13, 141-144, 150-152, 154-155, 162, 176, 180, 183, 229, 232-235, 249, 253
-Water · 53-54, 101, 106-107, 191, 209, 213
Demographic · 37, 88, 130, 152-153, 179, 183-184, 231, 233
Demonstration garden · *see* Garden, Demonstration
Depreciation · 65, 129
Desertification · 11, 247
Development pressure · 13, 32, 34, 62, 235
Digital aerial photo/analysis · 58, 151, 165, 178-180, 182-183 193-194

Diminishing returns · 66, 252
Direct costs · 72, 80, 108, 199, 248
Discounting rate · 127
Disease, plant (*including* Disease control) · 17-18, 24, 26, 35-36, 54, 56, 72-74, 91, 95, 98, 177
Display garden · *see* Garden, Display
Disposal of waste · *see* Waste Disposal
Domestic garden *see* Garden, Family/Home
Drought (*including* Dry period) · 53-54, 76, 79, 148, 167, 170, 185, 191, 213
-Hidden · 54, 251

E

Ecology/Ecological
-Benefits · 32, 87, 123, 158, 160, 217, 219
-Costs · 2, 9, 160-161, 251
-Farming · 44, 248
-Footprint · 15, 29, 101, 104, 107, 142, 199
-Function · 43-44, 116-118, 157, 194-195, 199, 217
-Goods and services (*also see* Environmental goods and services) · 38, 42, 44, 57, 59, 72, 114, 128, 199, 248, 251
-Gradients · 56-59, 247
-Impacts · 22, 40, 42, 63, 104, 114, 123
-Processes/Factors · 3, 23, 67, 118, 158, 161, 195, 245
-Sustainability · 22, 24, 123
-Systems · 18, 59
Economic(s) · 1-3, 5, 9-10, 15, 17, 19, 31-35, 37, 42, 44, 61-62, 64-65, 67-68, 87-89, 93, 122-124, 129, 133, 139, 149, 158, 219-224, 229, 237, 242-243, 248, 252
-Development · 16, 30
-Diversification · 123
-Forces · 3, 61, 64, 139, 209, 219
-Injury level · 78, 248
-Sustainability · 37-41, 65, 67, 98, 121, 123-124, 138, 140, 154, 157, 159-161, 209
Ecosystems · 6, 13, 17, 23, 39, 41-42, 44, 55, 59, 74, 110, 117-118, 138, 144, 246
Edible landscaping · 32
Educational garden · *see* Garden, Educational
Education-based farm · 69
Effects (*also see* Synergistic effects) · 63, 87-88, 90, 94, 100-101, 103, 108-110, 113-114, 117, 124-125, 127-129, 158, 197, 209, 214-215, 217, 220-229, 237, 240, 245, 247-248, 251-253, 255

Electricity · 77, 86, 105, 108-110, 116, 133, 172-173, 177, 191, 202-203, 212, 229

Elevation · 24, 48-51, 59, 251

Emissions · 12, 14-15, 26, 40, 43, 59, 80, 101, 103, 107-114, 116, 121, 128, 134-135, 161, 163, 172-174, 200, 203, 216-218, 225-226, 229, 246, 248, 250
-Methane · 100, 112-113, 215, 225-226
-Nitrous oxide · 111, 114

Energy · 11, 15-16, 23, 34, 43, 66, 71, 75, 84-85, 103, 107-111, 114, 116, 123, 125, 134-135, 142, 154, 161, 163, 171-173, 193, 200-202, 204, 212, 214-215, 218, 222-225, 227, 229, 241
- Non-renewable · 14-15, 20, 26, 43, 105, 108-110, 114, 200, 218

Environment/Environmental
-Considerations (*including* Issues, Concerns) · 11, 13, 23, 28, 32, 37-38, 40, 73, 85, 108, 121, 125, 139, 157-158, 167, 186, 193, 199, 209-210, 237, 246-7, 249
-Costs (*including* Degradation, Impacts) · 1, 6, 9, 11, 13,-14, 19, 29, 31, 44, 63, 72, 80, 103-105, 107-108, 111, 113-115, 118-119, 125, 127-128, 134, 160, 175, 200-204, 210, 213-218, 225, 244-245, 248, 251
-Goods and services (*including* Benefits) · 7, 35
-Gradients · 56, 59
-Parameters (*including* Factors) · 3, 45-47, 49-50, 67, 103, 170
-Protection/Restoration · 17, 24, 42, 87, 103-104, 118
-Quality · 26
-Resources · 41, 103
-Sustainability · 2, 37-39, 41, 98, 119, 123, 138, 140, 142, 154, 160-161, 209, 237, 242

Essential
-Activities · 71-72, 210, 239, 242
-Costs · 63, 72
-Plant nutrients · 55, 215
-Resources · 6, 36, 42, 74, 85, 103, 174, 211

Eutrophication · 103, 114

Evaporation · 48, 77, 104, 251, 253
-Evapotranspiration · 54, 77, 251
-Pan · 46
-Potential · 53-54

Exclusion bias · 129, 248

Expenditure · 61, 63-64, 81, 83-85, 105, 122-123, 133-134, 175, 177, 192, 210, 219-220

Expense · 2, 62-66, 68-69, 72-81, 84, 104, 121, 127-128, 131-134, 192, 210, 211, 220, 227, 248, 251, 254

Exposure · 35, 47-51, 57, 59, 73, 77, 109, 111, 129, 249

Extended costs · 72, 116, 129, 199, 249

External factor · 65, 237, 249

Extreme heterogeneity · 35, 58, 249

F

Family garden, *see* Garden, Family/Home

Farm
-Commercial · *see* Commercial, Farm
-Economics · *see* Economics, Farm
-Education-based · 69, 97
-Income · 65
-Market · 20, 29
-Non-commercial · *see* Non-commercial, Farm *or* Subsistence, Farm
-Runoff · *see* Runoff
-Size · 29, 62
-Small-scale · *see* Agriculture, Small-Scale
-Subsistence · *see* Subsistence Farm
-Truck · *see* Truck Farm
-Urban · *see* Urban, Farm
-Very small-scale · *see* Agriculture, Very Small-Scale Farm

Fertility · 18, 24, 42, 55-56, 77

Fertilizer · 17-18, 22-27, 40, 63, 65-66, 68, 72, 76-78, 80-82, 84, 90, 103, 108, 111, 113-115, 122-3, 163, 172, 174-175, 177, 192-193, 200-201, 213, 215-216, 218, 225-227, 237, 239, 245, 248, 251-253
-Chemical · 18, 24-26, 68, 108, 111, 113-115, 175, 177, 215-216, 225, 245, 251-252
-Natural/Organic · 22-24, 63, 80-81, 115, 163, 174-175, 192-193, 200-201, 218, 253

Financial investment · *see* Investment, Financial

Fixed costs · 65-66, 220, 249

Food demand (FD) · *see* Food, Demand
-Potential (FD$_P$) · *see* Food, Demand

Food
-And culture · 1, 17, 28, 31, 36-37, 61, 70, 87, 90-91, 94-95, 97, 118, 128, 152, 154, 159, 179, 205-208, 231, 238-240
-Definition · 137
-Demand · 1, 6, 11, 13, 15, 32-34, 39, 67, 70, 92, 141-144, 146, 150, 153, 155, 162, 180, 183, 232-234, 249, 253

-Deserts · 14, 33, 69, 91, 98, 123, 232, 238, 249
-Garden · 2-3, 11, 15, 21, 24, 30, 32-35, 43, 61-62, 64, 67, 70, 81, 84, 91, 94, 97, 125, 131-132, 143, 149, 157, 160-162, 169, 173-175, 177-190, 187, 189-194, 197, 201, 209, 211, 233
-Insecurity · 14, 33, 35, 98, 123, 249
-Plants · 15, 17-18, 27, 31, 49, 51, 65, 88-89, 91, 96-97, 117-118, 169, 178, 188, 255
-Production/growing/gardening · 1-2, 4, 6, 9-10, 13-16, 18-22, 24-27, 29, 31-37, 39-41, 43-45, 48-51, 54, 56-57, 61-63, 65-67, 69, 71-73, 76, 78-82, 85, 87-90, 94, 98, 100, 103-113, 115-119, 128, 143-145, 147, 150-152, 155, 161-162, 169, 171, 178-180, 183, 199, 233, 240-243, 249
-Production, historic (FP$_H$) · 145-146, 235, 251
-Production, potential (FP$_P$) · 145, 150-152, 183, 253
-Safety · 12, 24, 69, 98, 103
-Security · 14, 16, 37, 125, 138-139, 232, 238
-Sustainability · 2-4, 6, 41, 44, 137-144, 146, 150, 153-155, 159-163, 180, 210, 226, 229-235, 238-240, 243, 250
-System · 1-3, 7, 14-16, 18, 20, 25, 31-34, 39-40, 42, 68, 70, 107, 111, 123, 125, 138-139, 143-144, 153, 202, 217, 235, 242-244
Food sustainability index (FS$_I$) · 141-142, 154-155, 162, 180, 234-235, 250
Food sustainability measure (FS$_M$) · 141-142, 144, 153-154, 180, 233, 235, 250
Foodshed · 137-138, 141-148, 150-151, 153-155, 180, 249-250
-Efficiency · 144
Fossil fuel · see Fuel, Fossil
Frost · 49, 51-52, 79, 185, 192, 200, 204, 217-218
Fuel · 16-17, 29, 108-110, 116, 125, 172-173, 223, 243, 245-246
-Diesel · 107, 108, 172-173, 212, 217, 226, 246, 248
-Fossil · 43, 80, 100, 107-111, 114, 127, 158, 172-173, 200, 202-203, 212-213, 216-218, 222, 225-227, 229, 246, 248, 250
-Gasoline · 100, 108-111, 173, 217, 222, 226, 246, 250
Fungiculture · 17

G

Garden
-Backyard vegetable *see* Garden, Family/Home
-Community (*also see* Community, Garden/Gardening) · 11, 16, 19, 21, 25-26, 28-29, 31-34, 36-37, 43-44, 62, 64, 67, 69, 77, 81, 87, 91, 93-94, 97, 129, 131-132, 138, 143, 149-150, 160-161, 229, 234, 238, 242, 246
-Container · 26, 30, 133, 171, 192, 200, 218
-Demonstration · 29, 70, 133
-Display · 70
-Domestic · *see* Garden, Family/Home
-Educational · 15, 21
-Family/Home · 2-3, 11, 19, 21, 25-26, 30, 32, 34, 43, 62-64, 67, 70, 73, 75, 77, 81, 84, 87, 89, 91, 93-96, 125, 132, 142-143, 149-150, 157-162, 169, 173, 175, 178-180, 183, 189, 192-194, 197, 199, 202, 205, 210, 226, 230, 232-233, 235, 239
-Guerilla · 21, 251
-Niche · 70,
-Private · 32-33, 99
-Public · 32
-Rooftop · 16, 26
-School · 11, 33, 98
-Subsistence · 21
-Vertical gardens · 26, 133, 169, 171, 200, 216-218, 255
-Victory Gardens · 32-33, 235
Gardening skills · 36, 72, 210, 220
Gasoline · *see* Fuel, Gasoline
Genetically modified organisms · 18, 149
Governmental policy (*also see* Policy) · 20, 39, 54, 109, 125, 159, 217, 237-243, 249
Green roofs · 32
Greenhouse
-Gas damages · 109, 127
-Gas emissions · *see* Emissions
-Gasses · *see* Emissions
Greenspace · 15-16, 34-35, 56-59, 94, 101, 104, 107, 116-117, 145, 159, 178, 182, 193-194, 197, 199, 217, 235, 247, 250
Gross
-Benefits · 11, 138, 251
-Costs · 11, 251
Groundwater · 14, 43, 53, 85, 104-105, 125, 133, 171, 185, 200, 213-214, 218
-Pollution · 40, 101, 113, 213, 225-227

-Recharge · 35, 44, 101, 106-107, 115, 117, 193, 213
Growing
  -Conditions · 18, 34-36, 47, 49, 51-52, 58, 65, 83, 149, 160, 169, 180, 220
  -Method · 2, 10-11, 19, 22, 24, 26-27, 43-44, 62, 65, 75, 80, 103, 118, 132-133, 139, 154, 197, 219, 238, 243
  -Season · 27, 45-50, 66-67, 77, 81, 83-85, 107, 144, 146, 148, 150, 167-169, 183, 186-187, 189, 251
Guerilla garden · *see* Garden, Guerilla

H

Habitat · 11, 16, 57, 59, 72, 101, 116-117, 161, 169, 197, 219, 225
  -Impacts/Loss · 2, 7, 14, 26, 31, 37, 43, 72, 108, 112, 114, 116, 139, 175, 199, 200, 216, 218-219, 251
Harvest · 9, 18, 23, 27-29, 31, 36, 40, 46, 52-53, 64-66, 68, 71-75, 77-79, 82, 87-89, 101, 138-139, 149, 159, 172, 176, 187-190, 202, 210-211, 213, 230-231, 237, 245, 251
  -Value/valuation · 91-93, 115, 172, 175-176, 187, 189-190, 202, 212, 230-231, 246-247
Heat island effect · *see* Urban, Heat Island Effect
Heritage · 29, 39, 68, 70, 87, 94-97, 123, 137, 159, 179, 205, 207, 230-231
Hidden drought · 54, 251
High-density growing methods · 27, 65
High-intensity agriculture · 12, 25
Historic food production · *see* Production, historic
Home
  -Food gardens/growing · *see* Garden, Family/Home
Human
  -Diet/Dietary needs (*also* see Choose My Plate) · 41, 44, 91, 143-144, 146, 151, 209
Humidity · 46, 48, 53
Hydrogeology · 125
Hydroponic systems · 21-22, 25-26, 45, 75-76, 84-85, 104, 133, 251

I

Inca · 23-24
Income · 61-62, 64-65, 67-68, 121, 121, 221
Indirect
  -Benefits· 101, 121, 251
  -Costs · 72, 80-81, 100-101, 105, 108, 112-113, 115, 127, 194, 200, 217-218, 251
Industrial
  -Agriculture (also see Agriculture, Conventional; Commercial Agriculture) · 4, 7, 9-10, 12, 20-22, 25-26, 33, 36, 79, 104, 117, 243-244, 247, 251
  -Revolution · 23
Inflation · 61, 127, 190, 219, 221, 252
Informal interviews · *see* Interview, Informal
infrastructure · 15-16, 57, 68, 71-73, 74-75, 77, 81, 86, 90, 104, 109, 133, 177, 192, 254-255
Inputs · 2, 10, 13, 19, 26, 40, 61-62, 66, 71-72, 80, 82, 103, 113, 118, 121, 123, 125-129, 131, 133, 157, 159, 174, 201, 210-212, 247-248, 251-253
  -Intangible · 72, 87, 252
  -Tangible · 72, 87
  -Variable · 65-66, 82, 255
Intangible · 61, 74, 99-100, 121, 123-124, 161, 221, 223
  -Benefits (*including* Returns) · 10, 19, 72, 87-88, 94-97, 99-100, 119, 123, 127, 129, 132-133, 158, 161, 178, 205, 207-208, 251
  -Costs · 19, 79-80, 121, 123, 127, 207-208, 252
  -Inputs · 72, 74
Internal
  -Factors (*including* Effects, Processes) · 40, 65, 87, 94-95, 127, 210, 220-221, 223, 226, 252
  -Rate of return (IRR) · 127
Interview
  -Informal · 99
  -Semistructured · 99
  -Structured · 99
  -Unstructured · 99
Investment · 14, 16, 33, 62, 66-67, 74-76, 80-83, 90, 90, 106, 122, 125, 133, 157, 164, 192, 219, 227, 229-230, 247-248, 251-252, 255
  -Financial · 19, 62, 80, 82, 84, 87, 126-127, 163, 193, 210, 212, 220, 248
  -Labor/time · 99
Irrigation · 18, 20, 23-24, 30, 36, 43, 45, 53, 55-56, 64-66, 72, 74, 77, 79, 82, 84-85, 101,

104-105, 107-108, 111, 127, 133, 170-172, 185, 190-191, 200-201, 211-213, 216, 218, 222, 227, 237, 253

K

Kogi · 23-24

L

Labor · 18-20, 27, 29, 34, 41, 61-63, 65-66, 68, 72-76, 78-84, 114, 116, 123, 128-129, 132, 159, 163, 175, 177, 180, 191-192, 210-212, 220, 223, 226, 248, 252, 254
Land use · 32, 36, 101, 106, 134, 148, 164, 180, 241
Landfill · 15, 42, 100, 112-113, 118, 174, 201-202, 214-215, 225
Landscape · 12, 24, 35, 46, 48, 53, 56-57, 94, 97, 104, 109, 151, 153, 161, 173-174, 177, 212, 226, 240, 247
Large-scale agriculture · *see* Agriculture, Large-Scale
Latitude · 48-51, 167, 169, 251
Lawn · (*also see* Turfgrass) *34*, 106, 111, 117, 134, 169, 173, 177-178, 182, 194, 211, 213, 226-227, 250
Life cycle assessment · 85, 157
Light intensity · 35, 50
Limiting factors · 48, 77, 125
Local
    -Agriculture · 9, 15, 36, 107, 144-145, 153, 189, 221
    -Ecology · 10, 23, 103, 117-118, 255
    -Economy · 101, 210, 221
    -Food production ·15, 32, 34, 43, 98, 150, 152, 154, 180, 199, 221, 235, 239, 242
    -Food System · 16, 229

M

Macroeconomics · 61, 252
Market farm · *see* Farm, Market
Market/market system · 7, 11, 16, 20-21, 25-29, 32-34, 39, 41, 45, 52, 61, 64-65, 67-72, 74, 80, 82, 89-90, 93, 97, 108, 123-124, 127, 130-131, 134, 138-9, 143-144, 152, 160,

172, 176, 189-190, 205, 217, 221, 250, 252, 254-255
Methane · *see* Emissions, Methane
Microbial agriculture · *see* Agriculture, Microbial
Microclimate · 35, 57, 249, 252
Microeconomics · 61, 252
Migratory species · 101, 197-198, 217, 219, 225
Mining · 80, 114, 175, 215-216
Model · 32, 42, 99, 113, 122, 139, 153, 174, 215, 235, 237, 241, 253,
Moisture · 43, 47, 52-56, 76, 251
    -Availability · 48, 52, 54, 56, 76
    -Demand · 52-54
    -Gradient · 59, 246, 249
    -Regime · 54, 57, 59, 169
Monetary capital · 66, 87, 122, 125, 221
Monetary investment · *see* Financial Investment
Monetary-based project CBA · *see* Cost-Benefit Analysis, Monetary-based CBA
Monoculture · 26, 149, 251
Mulch · 82, 84, 115, 118, 171-174, 177, 192, 200-201, 218

N

Natural gas electricity generation · 109-10, 116, 172, 202, 248, 250
Natural
    -Habitat · *see* Habitat
    -Resources · 2, 11-13, 26, 40, 115, 123, 216
Net
    -Cash flow · 27
    -Present value (NPV) · 127, 219, 252
    -Value · 10, 127
New Urbanism · 241
Niche
    -Gardens · *see* Garden, Niche
    -Markets · 27, 254
NIMBY · 36, 252
Nitrates · 115
Nitrous oxide emissions · *see* Emissions, Nitrous Oxide
Non-climate damages · 173, 175
Non-commercial
    -Agriculture, Benefits of · 10-11, 19, 92-93, 94, 133-134
    -Crops · 27, 87, 89
    -Economics · 65, 67, 71, 87, 92-94, 134-135

-Farm (*including* Food-producing operations) · 61, 64, 67, 89, 93-94, 132, 160
-Farmer (*including* Food producers) · 25, 89, 91, 94, 132, 238
-Markets/market system · 93
-Urban agriculture · 94, 133, 157
Non-greenhouse gas impacts · 109, 111, 127, 172-173, 203, 223, 225
Non-monetized effects · 221, 223, 225
Non-profit farms · 62, 67-69, 252
Non-renewable energy · 14-15, 20, 26, 43, 105, 108, 134, 154
Non-renewable resources *see* Resources, Non-Renewable
Nuisance laws · 36, 241
Nutrient ·
-Cycling · 55
-Loss (*including* Runoff) · 114
-Plant · 9, 24, 44, 55, 78, 113, 175

O

On-going costs · 85, 105, 127
Operating costs · 14, 63, 66, 68, 85, 159, 253
Opportunity costs · 126-127, 226, 253
Orchards · 32
Organic
-Agriculture · 22-25, 28, 34, 43-44, 78, 115, 245, 253
- Fertilizers · 115, 174-5, 177, 192-3, 200, 218
- Matter (*including* Organic waste) · 12, 15, 63, 76, 101, 112-113, 115, 174, 201-202, 215, 222-223, 225
- Methods · 24-26, 34, 43, 63, 68, 89, 116, 176, 197, 217, 219
- Pest control · 172, 177, 192-193, 200, 218
- Soil · 42-44, 55-56, 59, 223
Organically grown · 25, 134, 176-177, 203-204
Organisms, beneficial · *see* Beneficial Organisms
Output(s) · 2, 10, 13, 25, 29, 40-41, 61-62, 67, 85, 91, 121, 123, 125-129, 131, 157, 170, 253-254

P

Pan evaporation · 46

Participant observation · 99
Permanent wilting point · 53
Pest
-Control (*also see* Organic Pest Control) · 17-18, 24, 31, 36, 43, 56, 64, 68, 78-79, 82, 84, 172, 177, 192-193, 200, 218, 227, 253
-Eradication · 56, 78
-Management · 18, 23-24, 78
Pesticides · 12, 20, 22-27, 43, 65, 78-79, 84, 108, 111, 117, 123, 226, 245, 251
- Chemical · 20, 111, 245, 251
Pests · 18, 23, 35-36, 43-44, 47, 55-56, 59, 71, 73-74, 76, 78, 82, 95, 195, 247, 254
Petroleum · 107-108, 116, 172, 217, 248, 250
Phosphate · 81, 113-114, 175
Photoperiodism · 49
Planning · 13-14, 17, 20, 54, 72-73, 81-82, 90, 214, 220-221, 229, 237, 239, 241-2
Plant
-Breeding · 18, 90
-Growth · 25, 27, 45-47, 49, 52-53, 76-78
-Productivity · (*see* Productivity)
-Respiration · 47
Policy (*also see* Governmental Policy) · 4, 9, 13, 20, 39, 54, 109, 121-122, 124-125, 129-131, 160, 217, 227, 229, 237, 239, 240-243, 249
-Resistance · 240
Political resistance · 238, 243
Pollinators · 28, 35, 41, 59, 117, 194-197
Pollution
-Agricultural · 13-14, 19, 26, 31, 72, 101, 103, 113-115, 119, 123, 127-128, 175, 200
-Issues · 11-12, 32, 35, 43, 52, 59, 63, 95, 107-110, 119, 139, 154, 199, 210, 214-216, 218, 244
Potential
-Food growing space · 180, 182-183, 232, 234, 241
-Food production · 141-142, 145-146, 150-151, 162, 180, 233, 249, 253
Poverty · 37
Precipitation (*also see* Rainfall) · 48, 53-54
Private gardens · *see* Garden, Private
Production costs · 64, 93, 190
Productivity · 45-47
-Agricultural · 10, 18, 20, 25, 31, 42, 44, 48, 51-52, 54-55, 65-66, 77-79, 88-91, 121, 137, 144, 146, 148-151, 154, 159, 162-163, 169, 175, 180, 183, 185, 187, 189, 211-212, 230-231, 234, 248, 253

Profit · 20-22, 28, 34-35, 39, 61-64, 67-69, 73-74, 79, 89-91, 93, 125, 127, 129, 133-134, 143, 148, 160, 220-221, 223

Profitability · 9, 19-21, 25, 32-33, 39, 61-62, 64-65, 80, 89, 121-122, 129, 144, 159, 161, 220, 223, 229

Proxy data · *see* Plant Proxy Data

Psychological benefits · 94, 97, 123

Public gardens · *see* Gardens, Public

Q

Quantitative methods · 41, 99

Questionnaires · 99-100, 179, 205-207, 240

R

Rain barrels · 171

Rainfall · 46-47, 53-55, 57, 65, 77, 79, 85, 106-107, 167-168, 170-171, 185-186, 245, 255

Rainwater · 85-86, 106-107, 133, 170-171, 191-192, 200-201, 216-218, 222

Raw materials · 11, 116, 123, 200, 214, 216, 218

Recycling of materials · *see* Waste, Recycling and Assimilation

Referent group · 124-125, 253

Renewable resource · *see* Resources, Renewable

Residential food garden · *see* Garden, Family/Home

Residential food garden case study ·

Resilience · 39-40, 230, 232, 243

Resources ·
-Conservation · 11, 15-16, 24, 39, 43, 54, 56, 104, 107, 113, 118, 161, 193, 199, 227, 229
-Energy · *see* Energy
-Limitations · 41
-Natural · *see* Natural Resources
-Non-renewable · 14, 43, 63, 139
-Renewable · 42-43
-Scarce · 121-122, 253
-Water · *see* Water

Retail price · 93, 176, 187

Returns · *see* Benefits; Commercial Agriculture, Benefits; Ecology/Ecological Benefits; Ecology/Ecological Goods and Services; Gross Benefits; Indirect Benefits; Intangible Benefits; Non-Commercial Agriculture,

Benefits of; Profit; Tangible Benefits; Tangible Effects

Risk · 5, 45, 49, 65-66, 78-79, 101, 106, 109, 124, 126-127, 129-130, 134, 175, 197, 200, 207, 213-214, 218, 225-227
-Management · 129

Rooftop gardens · *see* Garden, Rooftop

Runoff · 13-14, 26, 31, 40, 43, 53-54, 56-57, 85, 103, 106-107, 114-115, 127, 171, 175, 213

Rural food production · 20, 32

S

Sales · 20-21, 25, 28, 33, 61-62, 64-65, 68-69, 82, 89, 91, 93, 108, 121, 134, 152, 172, 246, 254

Salinization · 12, 253

Saltwater intrusion · 104, 106, 134, 200, 213-214, 218, 255-257

Scarce resources · *see* Resources, Scarce

School gardens · *see* Gardens, School

Seasons (*also see* Growing Season) · 9, 18, 25, 27-28, 46-50, 53, 63-64, 66-68, 74, 77-78, 80-85, 92, 104, 107, 144, 146, 148, 150, 167-169, 171, 174, 176, 183, 185-187, 189-171, 197, 201-202, 211-212, 215-217, 238, 251

Secondary impacts · 40

Seeds · 9, 33, 63, 68, 72, 74, 76, 82, 84, 123, 177, 192-193, 200, 218

Semistructured interviews · *see* Interviews, Semistructured

Sensitivity analysis · 227, 229, 253

Shade · 57, 77, 169

Small-scale agriculture/farm · *see* Agriculture, Small-Scale

Smart Growth · 241

Society/social considerations · 1-2, 4-6, 9-10, 13, 23, 26, 30, 32, 35-36, 42-43, 45, 61, 67, 69-70, 72, 87, 91, 95-99, 110, 121-125, 127, 130-131, 139-140, 143, 158, 160-161, 163, 178, 219, 237, 242-243, 250
-Sustainability · 22, 37, 41, 119, 208-209, 242

Soil (*also see* Organic Soil; Topsoil; Water Holding Capacity of Soil) · 17-18, 26, 40-43, 47, 54-55, 59, 65, 71-75, 76-77, 82, 106, 117, 123, 164, 177, 191-192, 252, 254
-Amendments · 42, 44-45, 76, 82, 84, 113, 115, 174, 177, 192-193, 200-202, 215-216, 218, 223

-Degradation · 12, 14, 26, 42, 56, 66, 103, 108, 253
-Environment/quality · 18, 24, 36, 42, 53-55, 59, 76-77, 90, 104, 113, 115, 118, 123, 247
Solar radiation · 41, 46, 50, 77
Specialty crop · 27, 89, 254
Stakeholders · 9, 37, 124-125, 254
Startup costs · 80-81, 127, 192
Statistics · 47, 49-51, 129, 152, 160, 170
Structured interviews · *see* Interviews, Structured
Subsistence
-Crops · 27-28, 254
-Farming · 19, 21-22, 27, 68, 70
-Farm · 23, 29, 62, 64, 67-69, 91, 132, 231
-Gardens · *see* Garden, Subsistence
Sunlight · *see* Solar Radiation
Suppliers · 20, 65, 93, 135, 221
Supply and demand · 64, 154
Sustainability
-Community · 14, 95, 159, 209
-Initiatives · 10, 14-16, 35, 125
Sustainable
-Agriculture · 3, 31, 39-40, 44, 138-140, 229, 237-238, 242-243
-Development · 11, 37
-Food condition · 154
Sustainably produced food · 139-140, 155, 229, 249-250
Synergistic effects · 129

T

Tangible
-Benefits · 19, 97, 255
-Costs · 71, 79-80
-Effects (*including* Returns) · 19, 87-88, 92-94, 100, 119, 127, 221
-Inputs · 72
-Products · 19, 248
Temperature · 35, 46-52, 57, 59, 79, 169-170, 185-186, 246, 249, 254-255
Topsoil · 55
Toxic waste · *see* Waste, Toxic
Traditional agriculture · 22-24, 27, 78, 255
Traditional crops · 27, 254-255
Truck farm · 20
Turfgrass (*also see* Lawn) · 111, 194, 213, 217, 226

U

Uncultivated plants (*also* see Weeds) · 194
United Nations Millennium Declaration · 11
Unstructured interviews · *see* Interviews, Unstructured
Unsustainable practices · 26, 138
Unusual crops · 28, 255
Urban
-Agriculture, commercial (*also* see Commercial Agriculture) · 10, 20, 32, 62, 65
-Ecology · 35, 43, 101, 103, 116-118, 163, 193, 210
-Environment (*including* Landscape *and* Setting) · 15, 22, 28, 32, 34, 47, 56-58, 117, 149, 177, 217, 240, 247
-Farm (*also see* Farm) · 32, 34, 36, 61-62, 65, 67, 150, 159
-Food gardens (*also see* Garden) · 15, 32-33, 43, 117, 154, 194, 235
-Food growers · 33, 36, 159, 161
-Food systems · 9, 17, 32, 123, 235
-Greenspace (*also see* Greenspace) · 16, 59, 117, 178-179, 193, 197, 235, 250
-Habitats · 15, 16, 59, 101
-Heat island effect · 35, 57, 193, 255
-Planning and policy · 13, 14, 17, 20, 121, 238-239
-Sprawl and development · 12-13, 32, 36, 57, 106, 178-179, 199, 241
-Sustainability · 10, 15, 31-32, 35, 39, 139, 163, 179, 199, 243

V

Valuation
-Methods · 92, 118, 122, 130, 134, 176, 199, 211
-Theory · 123, 128, 130, 255
Variable costs · 66, 219
Variable inputs · *see* Inputs, Variable
Vertical gardens · *see* Garden, Vertical
Very small-scale agriculture/farm · *see* Agriculture, Very Small-Scale
Victory Gardens · 32-33, 235, 255
Visible costs · 72

W

Waste · 15, 36, 111-112, 142, 152, 174, 200-201, 214-215, 218
   -Disposal · 75, 85, 100-101, 105, 108, 111-113, 118, 123, 127, 139, 174, 199-201, 214-215, 218, 225, 250
   -Products · 68, 103-105, 113-116, 118, 161, 163, 173, 201, 214-215, 218, 222-223, 225
   -Recycling and assimilation · 16, 42-43, 55, 63, 101, 103-104, 113, 115, 163, 173-174, 201-202, 214-216, 222-223, 226
   -Toxic · 36
Wastewater · 16, 85
Water · 1, 16, 24, 39, 40-44, 52-54, 59, 71-72, 75-77, 80, 82, 85-86, 101, 122-123, 125, 133-134, 163, 193, 200, 209, 213, 216-218, 223, 225-226, 245, 248, 250, 253-254
   -Availability · 16, 18, 36, 53-54
   -Conservation and management · 15-16, 18, 24, 35, 101, 104-107, 117-118, 161, 199, 213, 227
   -Holding capacity of soil · 55-56, 76, 115
   -Irrigation · 66, 82, 85-86, 127, 170-171, 177, 185, 190-192, 201, 211-212, 216-218, 222-223

   -Quality · 11, 14, 23, 40, 43, 77, 101, 103, 108, 110, 113-115, 123, 213-214, 226-227, 253-254
   -Resources · 11, 23, 26, 35, 66, 101, 103-107, 125, 133-134, 163, 213
   -Sources · 15-16, 77, 85-86, 104-107, 118, 125, 133
   -Temple · 23
Weather · 18, 35, 45-51, 73, 78-79, 90, 148, 170, 185, 187, 191, 213, 220, 243, 246
Weeds (*also* see Uncultivated Plants) · 18, 35, 76, 79, 82, 89, 112, 117-118, 177, 194-195
Wholesale pricing · 93, 190
Wild foods · 15. 138
Wildlife · 96, 117, 161, 169-170, 177, 194-195, 199, 205, 217
Willingness to pay · 124, 128, 130-131, 211, 255
Wind · 35, 46, 48, 50, 52, 56-57, 59, 79, 109, 185, 249
Windbreaks · 52, 56
Working capital · 41, 127
World population · 2, 6, 11-13, 40, 116

Z

Zoning · 36, 240-242

References

Aiking, H. & J. De Boer. 2004. Food sustainability: Diverging interpretations. *British Food Journal*, *106*(5), pp.359-365.

Alaimo, K., E. Packnett, R.A. Miles & D.J. Kruger. 2008. Fruit and vegetable intake among urban community gardeners. *Journal of nutrition education and behavior*, *40*(2), pp.94-101.

Alder, M.D. & E.A. Posner. 2001. Cost-benefit Analysis: Economic, Philosophical and Legal Perspectives. University of Chicago Press.

Alexander, T. *et al.* 1999. The best of growing edge. New Moon Publishing.

Altieri, M.A., N. Companioni, K. Cañizares, C. Murphy, P. Rosset, M. Bourque & C.I. Nicholls. 1999. The greening of the "barrios": Urban agriculture for food security in Cuba. *Agriculture and Human Values*, *16*(2), pp.131-140.

American Community Gardening Association. 2016. Website information available at https://communitygarden.org; last accessed May 22, 2017.

American Transport Research Institute. 2009. Website information available at http://www.maine.gov/mdot/ofbs/docs/atrimainereport.pdf; last accessed May 22, 2017.

Assuncao, J.J. & M. Ghatak. 2003. Can unobserved heterogeneity in farmer ability explain the inverse relationship between farm size and productivity. *Economics Letters*, *80*(2), pp.189-194.

Aubry, C., J. Ramamonjisoa, M.H. Dabat, J. Rakotoarisoa, J. Rakotondraibe & L. Rabeharisoa. 2012. Urban agriculture and land use in cities: An approach with the multi-functionality and sustainability concepts in the case of Antananarivo (Madagascar). *Land Use Policy*, *29*(2), pp.429-439.

Australian Bureau of Statistics. 1992. *Home production of selected foodstuffs, Australia, year ended in April 1992*. Catalogue No. 71 10.0, ABS, Canberra, ACT.

Kawane, A., T. Aikoh & S. Asakawa. 2000. Attitudes of the residents toward private gardening in Megumino, Hokkaido. *Journal of the Japanese Institute of Landscape Architecture*, *63*(5), pp.695-700.

Bahnson, F. 2010. Organic by Necessity: Agriculture in Havana. *The Christian Century*, *127*(18), pp.11-12.

Barrett, C.B., 1996. On price risk and the inverse farm size-productivity relationship. *Journal of Development Economics*, *51*(2), pp.193-215.

Beech, B.M., R. Rice, L. Myers, C. Johnson & T.A. Nicklas. 1999. Knowledge, attitudes, and practices related to fruit and vegetable consumption of high school students. *Journal of Adolescent Health*, *24*(4), pp.244-250.

Berke, P.R., J. MacDonald, N. White, M. Holmes, D. Line, K. Oury & R. Ryznar. 2003. Greening development to protect watersheds: Does new urbanism make a difference? *Journal of the American Planning Association, 69*(4), pp.397-413.

Bernard, H.R. 2011. Research methods in anthropology: Qualitative and quantitative approaches. Rowman Altamira.

Bissett, T.L. 1976. Community Gardening in America. *Brooklyn Botanical Garden Record, 35*, 4.

Blair, D. 2009. The child in the garden: An evaluative review of the benefits of school gardening. *The Journal of Environmental Education, 40*(2), pp.15-38.

Bliatout, B.T. 1986. Guidelines for mental health professionals to help Hmong clients seek traditional healing treatment. *The Hmong in transition*, pp.349-363.

Bouwman, A.F. 1996. Direct emission of nitrous oxide from agricultural soils. *Nutrient cycling in agroecosystems, 46*(1), pp.53-70.

British Columbia Ministry of Agriculture & Lands. 2006. B.C.'s Food Self-Reliance; Can B.C.'s Farmers Feed Our Growing Population? Website information available at http://www2.gov.bc.ca/assets/gov/farming-natural-resources-and-industry/agriculture-and-seafood/agricultural-land-and environment/ strengthening-farming/800-series/820105-1_bcfoodselfreliance_report.pdf; last accessed May 22, 2017.

Broadway, M. 2009. Growing Urban Agriculture in North American Cities: The Example of Milwaukee. *Focus on Geography, 52*(3-4), pp.23-30.

Brown, D.K. 2004. Cost-benefit analysis in criminal law. *California Law Review*, pp.323-372.

Brown, K.H. & A.L. Jameton. 2000. Public health implications of urban agriculture. *Journal of Public Health Policy*, pp.20-39.

Brundtland, G.H. & M. Khalid. 1987. Our common future. *New York*.

Burton, P. 2012. Grow your own: making Australian cities more food-secure. Website information available at http://theconversation.com/grow-your-own-making-australian-cities-more-food-secure-8021; last accessed May 22, 2017.

Byers, K. 2009. The psycho-social benefits of backyard and community gardening among immigrants. Masters, University of Georgia.

Campbell, H.F. & R.P. Brown. 2003. Benefit-cost analysis: financial and economic appraisal using spreadsheets. Cambridge University Press.

Carmichael, M.C. 2011. Putting down roots; gardening insights from Wisconsin's early settlers. Wisconsin Historical Society Press.

Carter, M.R. 1984. Identification of the Inverse Relationship between Farm Size and Productivity: An Empirical Analysis of Peasant Agricultural Production. *Oxford Economic Papers, New Series 36*(1):131-145.

References

Clayton, S. 2007. Domesticated nature: Motivations for gardening and perceptions of environmental impact. *Journal of environmental psychology*, *27*(3), pp.215-224.

Cobb, T.D. 2012. Reclaiming our food: How the grassroots food movement is changing the way we eat. Storey Publishing.

Collum, P.J. 1995. *Production of vegetable crops.* APS Press.

Colt, J. 2011. Canary in a phosphate mine. June 17, 2011 CNN special report. Document can be accessed at http://www.cnn.com/2011/IREPORT/06/17/nauru.colt/index.html; last accessed on May 22, 2017.

Corrigan, M.P. 2011. Growing what you eat: Developing community gardens in Baltimore, Maryland. *Applied Geography, 31*, pp.1232-1241.

Costanza, R. *et al.* 1997. The value of the world's ecosystem services and natural capital. *Nature, 387*(6630), pp.253-260.

Cranes Detroit Business. 2016. Detroit plots course on land use a leader in urban farming. August 6, 2016 issue; article can be accessed at http://crainsdetroit.com/article/20160806/NEWS/160809907/detroit-plots-course-on-land-use-as-a-leader-in-urban-farming; last accessed on May 22, 2017.

Daily Detroit. 2015. Detroit urban farms rooting goodness into the city. July 6, 2015 issue; article can be accessed at http://www.dailydetroit.com/2015/07/06/10-detroit-urban-farms-rooting-goodness-into-the-city/; last accessed on May 22, 2017.

Daniels, T.L. 2008. *Farmland preservation in growth management: Lessons for Florida.* Florida State University DeVoe Moore Center Critital Issues Symposium 2008.

Daniels, T.L. 2009. A trail across time: American environmental planning from city beautiful to sustainability. *Journal of the American Planning Association, 75*(2), pp.178-192.

Debertin, D.L. 2012. Agricultural production economics. CreateSpace Independent Publishing Platform.

DeLind, L.B. 2002. Place, work and civic agriculture: Common fields for cultivation. *Agriculture and Human Values, 19*(3), pp.217-224.

De Vries, S., R.A. Verheij, P.P. Groenewegen & P. Spreeuwenberg. 2003. Natural environments-healthy environments? An exploratory analysis of the relationship between greenspace and health. *Environment and planning A, 35*(10), pp.1717-1732.

Diekelmann, J. & R.M. Schuster. 2002. *Natural Landscaping: Designing With Native Plant Communities.* University of Wisconsin Press.

Downs, A. 2005. Smart growth; why we discuss it more than we do it. *Journal of the American Planning Association 71*(4), pp.367-380.

Dunnett, N. & M. Qasim. 2000. Perceived benefits to human well-being of urban gardens. *HortTechnology, 10*(1), pp.40-45.

Edwards, A.R. 2010. Thriving beyond sustainability; Pathways to a resilient society. New Society Publishers.

Elmqvist, T. 2012. Cities and biodiversity outlook-action and policy. In Montreal, Canada: UN Secretariat of the. Convention of Biological Diversity.

Environmental Defense Fund. 2017. Website information available at https://www.edf.org/blog/2014/05/07/trucks-delivering-six-miles-gallon-wont-work-long-haul; last accessed May 22, 2017.

Faist, M., S. Kytzia & P. Baccini. 2001. The impact of household food consumption on resources and energy management. *International Journal of Environment and Pollution, 15*(2), pp.183-199.

Fivush, R., J.G. Bohanek & M. Duke. 2005. The Intergenerational Self: Subjective Perspective and Family History. Individual and collective self-continuity. Erlbaum.

Florida Department of Environmental Protection. 2017. Phosphate mines. Website information available at https://www.dep.state.fl.us/water/mines/manpho.htm; last accessed May 22, 2017.

Florida Fish and Game Commission & The American Farmland Trust. 1995. A landowner's strategy for protecting Florida panther habitat on private lands in south Florida. American Farmland Trust.

Francis, C. & C. Marcus. 1991. Places people take their problems [Thematic Session Affective relations between 'person-object-place']. In: J. Urbina-Soria, P. Ortega-Andeane, & R. Bechtel (Eds.), *Healthy environments* [Proceedings of EDRA 22 / 1991] (pp. 178-184). EDRA

Francis, M. & R.T. Hester, Jr. 1990. *The meaning of gardens.* The Massachusetts Institute of Technology Press.

Freeman, C., K.J.M. Dickinson, S. Porter & Y. van Heezik. 2012. "My Garden is an Expression of Me": Exploring Householder's Relationships With Their Gardens. *Journal of Environmental Psychology, 32,* pp.135-143.

Fuguitt, D. & S.J. Wilcox. 1999. *Cost-Benefit Analysis for Public Sector Decision Makers.* Westport, Connecticut.

Fuller, R.J., *et al.* 2005. Benefits of organic farming to biodiversity vary among taxa. *Biology letters, 1*(4), pp.431-434.

Garnett, T. 1996. Harvesting the Cities. *Town and Country Planning, 65*(10), pp.264-266.

Garnett, T. 2013. Food sustainability: problems, perspectives and solutions. *Proceedings of the Nutrition Society, 72*(01), pp.29-39.

Gaston, K.J., P.H. Warren, K. Thompson & R.M. Smith. 2005. Urban domestic gardens (IV): The extent of the resource and associated features. *Biodiversity and Conservation, 14,* pp.3327-3349.

Getz, A. 1991. Urban foodsheds. *The Permaculture Activist, 24*(October), pp.26-27.

Ghosh, S. 2010. Sustainability potential of suburban gardens: Review and new directions. *Australasian Journal of Environmental Management, 17*(3), pp.165-175.

Ghosh, S., R. Vale & B. Vale. 2007. Metrics of local environmental sustainability: A case study in Auckland, New Zealand. *Local Environment, 12*(4), pp.355-378.

Ghosh, S., R. Vale & B. Vale. 2008. Local food production in home gardens: Measuring on-site sustainability potential of residential development. *International Journal of Environment and Sustainable Development, 7*(4), pp.430-451.

Gordon, P. & H.W. Richardson. 1997. Are compact cities a desirable planning goal? *Journal of the American Planning Association, 63*, pp.95-106.

Greenbiz. 2015. Urban farms now produce 1/5 of the world's food. May 5, 2015 issue; article can be accessed at https://www.greenbiz.com/article/urban-farms-now-produce-15-worlds-food; last accessed May 22, 2017.

González Jácome, A. 2009. Mexico: Traditional Agriculture as a Foundation for Sustainability. *From Traditional to Sustainable Agriculture*, pp.179-204.

Gu, C., J. Crane, G. Hornberger and A. Carrico. 2015. The effects of household management practices on the global warming potential of urban lawns. *Journal of environmental management, 151*, pp.233-242.

Hall, D. 1996. A garden of one's own: The ritual consolations of the backyard garden. *Journal of American Culture, 19*(3), pp.9-13.

Hedden, W.P. 1929. How great cities are fed. D.C. Heath and Co.

Helms, M. 2004. Food sustainability, food security and the environment. *British Food Journal, 106*(5), pp.380-387.

Hamner, K.C. & J. Bonner. 1938. Photoperiodism in relation to hormones as factors in floral initiation and development. *Botanical Gazette, 100*(2): pp.388–431.

Head, L. & P. Muir. 2007. *Backyard: Nature and culture in suburban Australia.* University of Wollongong Press.

Hepperly, P.R., D. Douds and R. Seidel. 2006. The Rodale Institute Farming Systems Trial 1981 to 2005: long-term analysis of organic and conventional maize and soybean cropping systems. *Long-term Field Experiments in Organic Farming. Berlin: Verlag Dr. Köster*, pp.15-31.

Hoffpauir, J. 2009. Environmental Impact of Commodity Subsides: NEPA and the Farm Bill. *The Fordham Environmental Law Review, 20*, pp.233.

Hole, D.G., A.J. Perkins, J.D. Wilson, I.H. Alexander, P.V. Grice & A.D. Evans. 2005. Does organic farming benefit biodiversity?. *Biological conservation, 122*(1), pp.113-130.

Holm, D.L. 2001. Massachusetts agriculture and food self-sufficiency: An analysis of change from 1974 through 1997. Electronic Doctoral Dissertations for UMass Amherst. Paper AAI3027207, document can be accessed at http://scholarworks.umass.edu/dissertations/AAI3027207; last accessed May 22, 2017.

Hunt Jr., J.B., G.J.B. Howes & D.P.P.E.A. Director. 1997. Analysis of the Full Costs of Solid Waste Management for North Carolina Local Governments. *DPPEA, February.*

Ilbery, B., Q. Chiotti & T. Rickard. 1997. *Agricultural Restructuring and Sustainability: A Geographical Perspective.* Centre for Agricultural Bioscience International.

Jones, M., N. Dailami, E. Weitkamp, D. Salmon, R. Kimberlee, A. Morley & J. Orme. 2012. Food sustainability education as a route to healthier eating: evaluation of a multi-component school programme in English primary schools. *Health education research, 27*(3), pp.448-458.

Kaplan, R. 1973. Some psychological benefits of gardening. *Environment and Behavior, 5,* pp.145-162.

Kates, R.W., T.M. Parris & A.A. Leiserowitz. 2005. What is sustainable development? *Environment, 47*(3), pp.8.

Kiesling, F.M. & C.M. Manning. 2010. How Green is Your Thumb? Environmental Gardening Identity and Ecological Gardening Practices. *Journal of Environmental Psychology, 30,* pp.315-327.

Kimber, C. 2004. Gardens and dwelling: People in vernacular gardens. *Geographical Review, 94*(3), pp.263-283.

Koch, S., T.M. Waliczek & J.M. Zajicek. 2006. The Effect of a summer garden program on the nutritional knowledge, attitudes, and behaviors of children. *HortTechnology, 16*(4), pp.620-625.

Kortright, R. 2007. *Edible backyards: Residential land use for food production in Toronto.* Master of Arts thesis, Department of Geography, Collaborative Program in Environment and Health, University of Toronto.

Lacy, P. 2006. *Farmland preservation: The benefits of saving our agricultural land and resources; Practice guide #16.* Center for Environmental Policy and Management, University of Louisville, Louisville, KY.

Lang, Tim, & D. Barling. 2010. Food security and food sustainability: reformulating the debate. *The Geographical Journal* 178.4 (2012): pp.313-326.

Lautenschlager, L. & C. Smith. 2006. Beliefs, knowledge and values held by inner-city youth about gardening, nutrition and cooking. *Agriculture and Human Values, 24,* pp.245-258.

Lee-Smith, D. 2010. Cities feeding people: An update on urban agriculture in equatorial Africa. *Environment and Urbanization, 22,* pp.483-499.

Lewis, P. 1993. The making of vernacular taste: The case of Sunset and Southern Living. The Vernacular Garden, ed. John Dixon Hunt and Joachim Wolschke-Bulmahn. Washington, DC: Dumbarton Oaks, pp.107-36.

References

Linebergert, S.E. & J.M. Zajicek. 2000. School gardens: Can a hands-on teaching tool affect students' attitudes and behaviors regarding fruit and vegetables? *HortTechnology, 10*(3), pp.593-597.

Loram, A., P. Warren, K. Thompson & K. Gaston. 2011. Urban domestic gardens: The effects of human interventions on garden composition. *Environmental Management, 48, pp.*808-824.

Lyson, T.A. & A. Guptill. 2004. Commodity Agriculture, Civic Agriculture and the Future of U.S. Farming. *Rural Sociology, 69*(3), pp.370-385.

Macher, R. 1999. Making your small farm profitable. Storey Publishing.

Marsden, T., J. Murdoch, P. Lowe, R. Munton & A. Flynn 1993. *Constructing the Countryside.* University College London Press.

Martin, R. & T. Marsden. 1999. Food for urban spaces: The development of urban food production in England and Wales. *International Planning Studies, 4*(3), pp.389-411.

Martinez, S. 2010. Local food systems; concepts, impacts, and issues. Diane Publishing.

McAleese, J.D. & L.L. Rankin. 2007. Garden-based nutrition education affects fruit and vegetable consumption in sixth-grade adolescents. *Journal of the American Dietetic Association, 107*(4), pp.622-665.

McConnel, V. & M. Walls. 2005. The value of open space: Evidence from studies of nonmarket benefits. Resources for the Future.

McCormack, L.A., M.N. Laska, N.I. Larson & M. Story. 2010. Review of the nutritional implications of farmer's markets and community gardens: A call for evaluation and research efforts. *Journal of the American Dietetic Association, 110*, pp.399-408.

McIlvaine-Newsad, H., C.D. Merrett, W. Maakestad & P. Mclaughlin. 2008. Slow food lessons in the fast food Midwest. *Southern Rural Sociology, 23*(1), pp.72-93.

Mendes, W., L. Balmer, T. Kaethler & A. Rhodes. 2008. Using land inventories to plan for urban agriculture; Experiences from Portland and Vancouver. *Journal of the American Planning Association, 74*(4), pp.435-449.

Minno, M.C. & M. Minno. 1999. Florida butterfly gardening: A complete guide to attracting, identifying and enjoying butterflies. University Press of Florida.

Møller, V. 2005. Attitudes to food gardening from a generational perspective. *Journal of Integrated Relationships, 3*(2), pp.63-80.

National Audubon Society. 2017. Atlantic Flyway. Website information available at http://www.audubon.org/sites/default/files/documents/ar2011-flywayconservation.pdf; last accessed May 22, 2017.

National research Council. 2010. Hidden costs of energy: unpriced consequences of energy production and use. National Academies Press.

National Resource Council. 2016. Website information available at http://www.nationalacademies.org/nasem/; last accessed May 22, 2017.

National Weather Service. 2010. 2010 South Florida Weather Year in Review. Document available at https://www.weather.gov/media/mfl/news/2010WxSummary.pdf; last accessed May 22, 2017.

Natural Resources Conservation Service. 2016. Cropland. Website information available at https://www.nrcs.usda.gov/wps/portal/nrcs/main/?ss=16&navtype=BROWSEBYSUBJECT&cid=null&navid=960100000000000&position=BROWSEBYSUBJECT&ttype=main; last accessed May 22, 2017.

Nemecek, T., A. Heil, O. Huguenin, S. Meier, S. Erzinger, S. Blaser, D. Dux & A. Zimmermann. 2007. Life cycle inventories of agricultural production systems. *Final report ecoinvent v2. 0 No, 15.*

Nichol, K.L. 2001. Cost-benefit analysis of a strategy to vaccinate healthy working adults against influenza. *Archives of Internal Medicine, 161*(5), pp.749.

Nielsen, T.S. & K.B. Hansen. 2007. Do Green Areas Affect Health? Results from a Danish Survey on the Use of Green Areas and Health Indicators. *Health & Place, 13*(4), pp.839-850.

Nijmeijer, M., W. Worsley & B. Astill. 2004. An exploration of the relationships between food lifestyle and vegetable consumption. *British Food Journal, 106*(7), pp.520-533.

Nugent, R.A. 1999. Measuring the sustainability of urban agriculture. *For hunger-proof cities. Sustainable urban food systems*, pp.95-99.

Office of Disease Prevention and Health Promotion. 2017. Website information available at https://health.gov/dietaryguidelines/purpose.asp; last accessed May 22, 2017.

Okvat, H.A. & A.J. Zautra. 2011. Community Gardening: A Parsimonious Path to Individual, Community and Environmental Resilience. *American Journal of Community Psychology, 47, pp.*374-387.

Orr, D.W. 2002. The nature of design; Ecology, culture and human intention. New York, NY: Oxford University Press.

Palm Beach Post. 2011. *Rain in Forecast, but Area to Stay Thirsty.* Article published March 27, 2011.

Peoples, M.B., D.F. Herridge & J.K. Ladha. 1995. Biological nitrogen fixation: an efficient source of nitrogen for sustainable agricultural production?. *Plant and soil, 174*(1-2), pp.3-28.

Peters, C.J., N.L. Bills, J.L. Wilkins & G.W. Fick. 2009a. Foodshed analysis and its relevance to sustainability. *Renewable Agriculture and Food Systems, 24*(01), pp.1-7.

Peters, C.J., N.L. Bills, A.J. Lembo, J.L. Wilkins & G.W. Fick. 2009b. Mapping potential foodsheds in New York State: A spatial model for evaluating the capacity to localize food production. *Renewable Agriculture and Food Systems*, *24*(01), pp.72-84.

Peters, K.A. 2010. Creating a Sustainable Urban Agriculture Revolution. *Journal of Environmental Law and Litigation, 25,* pp.203.

Peterson, A. 2000. Alternatives, Traditions, and Diversity in Agriculture. *Agriculture and Human values, 17*(1), pp.95-106.

Pimentel, D., P. Hepperly, J. Hanson, D. Douds & R. Seidel. 2005. Environmental, Energetic, and Economic Comparisons of Organic and Conventional Farming Systems. *BioScience 55* (7), pp.573-582.

Pothukuchi, K. 2004. Community food assessment: A first step in planning for community food security. *Journal of Planning Education and Research, 23,* pp.356-377.

Premanandh, J. 2011. Factors affecting food security and contribution of modern technologies in food sustainability. *Journal of the Science of Food and Agriculture, 91*(15), pp.2707-2714.

Purdue University. 2017. Cassava fact sheet. Document available at https://hort.purdue.edu/newcrop/CropFactSheets/cassava.html#Production%20Practices; last accessed May 22, 2017.

Randall, T.A., C.J. Churchill & B.W. Baetz. 2003. A GIS-based decision support system for neighbourhood greening. *Environment and Planning, 30,* pp.541-563.

Rees, W.E. 1992. Ecological footprints and appropriated carrying capacity: What urban economics leaves out. *Environment and Urbanization, 4*(2), pp.121-130.

Rees, W.E. & M. Wackernagel. 1996. Urban ecological footprints: Why cities cannot be sustainable- and why they are a key to sustainability. *Environmental Impact Assessment Review, 16,* pp.223-248.

Reganold, J.P., L.F. Elliott & Y.L. Unger. 1987. Long-term effects of organic and conventional farming on soil erosion. *Nature, 330*(6146), pp.370-372.

Relf, D., A.R. McDaniel & B. Butterfield. 1992. Attitudes toward plants and gardens. *HortTechnology, 2*(2), pp.201-204.

Roberts, C.E. 1992. *Textural analysis of urban thematic mapper data.* Doctoral dissertation, Pennsylvania State University.

Rubinstein, N.J. 1997. The psychological value of open space. In L. W. Hamilton (Ed.), *The benefits of open space* (Chapter 4). The Great Swamp Watershed Association.

Ryall, C. & P. Hatherell. 2003. A Survey of Strategies Adopted by UK Wildlife Trusts in Promotion of Gardening for Wildlife. *The Environmentalist, 23,* pp.81-87.

Ryan, R.M., N. Weinstein, J. Bernstein, K.W. Brown, L. Mistretta & M. Gagne 2010. Vitalizing Effects of Being Outdoors and in Nature. *Journal of Environmental Psychology, 30*(2), pp.159-168.

Senauer, B., E. Asp & J. Kinsey. 1991. The food industry: An overview and implications of consumer trends, in food trends and the changing consumer. Eagan Press.

Shutkin, W. A. 2000. The land that could be; Environmentalism and democracy in the Twenty-First Century. The Massachusetts Institute of Technology Press.

Sinclair, D. 2005. *The Spirituality of Gardening*. Northstone Publishers.

Skelly, S. M. & J.C. Bradley. 2000. The importance of school gardens as perceived by Florida Elementary School Teachers. *HortTechnology, 10*(1), pp.229-231.

Small, R. 2007. Organic Gardens bring hope to poor urban communities. *Appropriate Technology, 34*(1).

Smit, J. & J. Nasr. 1992. Urban agriculture for sustainable cities: Using wastes and idle land and water bodies as resources. *Environment and Urbanization, 4*(2), pp.141-152.

Sonntag, V. 2008. Why local linkages matter; Findings from the local food economy study. Sustainable Seattle.

South Florida Water Management District. 2009. Land Cover/Land Use database; 2008-2009 update. South Florida Water Management District, West Palm Beach, FL.

Sperling, C.D. & C.J. Lortie. 2010. The Importance of Urban Backgardens on Plant and Invertebrate Recruitment: A field Microcosm Experiment. *Urban Ecologist, 13*, pp.223-235.

Stone, Jr., B. & M.O. Rodgers. 2001. How the design of cities influences the urban heat island effect. *Journal of the American Planning Association, 67*(2), pp.186-198.

Subair, S. K. & M. Siyana. 2003. Attitude toward backyard gardening in Botswana. *Proceedings of the 19th Annual Conference AIAEE 2003*, Raleigh, NC.

Surbeck, A. & D. Leu. 1998. Toxikologische beurteilung von nitrat im trinkwasser. *Zürich, ETH-UNS-Fallstudie, 97*, pp.203-205.

Szlanfucht, D.L. 1999. How to save American's depleting supply of farmland. *Drake Journal of Agricultural Law, 4*, pp.333-355.

Thompson, Jr., E., A.M. Harper & S. Kraus. 2008. San Francisco Foodshed Assessment (2008). American Farmland Trust.

Tse, M.M.Y. 2010. Therapeutic Effects of an Indoor Gardening Programme for Older People Living in Nursing Homes. *Journal of Clinical Nursing, 19*(7-8), pp.949-958.

Union of Concerned Scientists USA. 2016. Subsidizing waste. Website information available at http://www.ucsusa.org/our-work/food-agriculture/advance-sustainable-agriculture/subsidizing-waste#.WD4FgBSU4mc; last accessed May 22, 2017.

United Nations. 2000. United Nations Millennium Declaration. Website information available at http://www.un.org/en/ga/search/view_doc.asp?symbol=A/55/L.2; last accessed May 22, 2017.

United Nations. 2012. World Development Indicators (WDI). Website information available at http://data.worldbank.org/data-catalog/world-development-indicators; last accessed May 22, 2017.

United States Agricultural Survey #1. 2012. Website information available at https://www.agcensus.usda.gov/Publications/2012/Full_Report/Volume_1,_Chapter_2_County_Level/Florida/; last accessed May 22, 2017.

United States Census Bureau. 2016. World population; total midyear population for the world: 1950-2050. Website information available at http://www.census.gov/population/international/data/worldpop/table_population.php; last accessed May 22, 2017.

United States Department of Agriculture. 1978. Status Report: Small Farms in the US. USDA. 05-01-1998.

U.S. Department of Agriculture. 1981. 1978 Census of Agriculture; State and County Data, Florida.

United States Department of Agriculture. 2010. Small farms in the United States, Persistence under pressure. Economic Research Service, Economic Information Bulletin No. 63, February 2010.

United States Department of Agriculture. 2016. Choose My Plate food consumption guidelines Website information available at http://www.choosemyplate.gov/vegetables; last accessed May 22, 2017.

United States Department of Agriculture Economic Research Service (ERS). 2016. Organic production. Website info available at https://www.ers.usda.gov/Data-products/organic-production.aspx; last accessed May 22, 2017.

United States Department of Agriculture Soil Conservation Service. 1978. Soil Survey of Palm Beach County Area, Florida.

U.S. Department of Energy. 2014. Alternatives Fuel Data Center- Fuel properties comparison. Data available at https://www.afdc.energy.gov/fuels/fuel_comparison_chart.pdf; last accessed May 22, 2017.

United States Department of Environmental Protection. 2016. Heat island effect. Website information available at http://www2.epa.gov/heat-islands; last accessed May 22, 2017.

United States Energy Information Administration. 2014. Frequently asked questions-how much carbon dioxide is produced by burning gasoline and diesel fuel? Website information available at https://nnsa.energy.gov/sites/default/files/nnsa/08-14-multiplefiles/DOE%202012.pdf; last accessed May 22, 2017.

United States Energy Information Administration. 2016. Frequently asked questions-how many gallons of gasoline and diesel fuel are made from one barrel of oil? Website information available at https://www.eia.gov/tools/faqs/faq.php?id=327&t=9; last accessed May 22, 2017.

United States Environmental Protection Agency. 1947. "Federal Insecticide, Fungicide, and Rodenticide Act (FIFRA), 1947." *US Environmental Protection Agency.*

United States Environmental Protection Agency. 2008. Overview of greenhouse gasses. Website information available at https://www.epa.gov/ghgemissions/overview-greenhouse-gases; last accessed May 22, 2017.

United States Environmental Protection Agency. 2014. Landfill gas Monte Carlo Model documentation and results. Website information available at https://19january2017snapshot.epa.gov/www3/warm/pdfs/lanfl_gas_mont_carlo_modl.pdf; last accessed May 22, 2017.

United States Geological Survey. Water questions and answers, how much water does the average person use at home per day? Website information available at https://water.usgs.gov/edu/qa-home-percapita.html; last accessed May 22, 2017.

United Nations. 2016. The world at six billion. Table 2. Department of Economic and Social Affairs, Population Division.

University of Illinois at Urbana-Champaign. 2013. Returns and cash rents given $4.80 corn and $10.75 soybean prices. FarmdocDaily, Department of Agriculture and Consumer Economics, July 16, 2013.

University of Missouri Cooperative Extension, *et al.* 1965. Farm Business Planning Guide for Organization #6500, University of Missouri Cooperative Extension, Columbia, MO, 1965; Selected Fruit and Vegetable Planning Budgets, EC 959, by Charles D. DeCourley and Kevin C. Moore, Department of Agricultural Economics, University of Missouri- Columbia, 1987; and Enterprise Budgets: Northeast Oklahoma 1985, prepared by Bill Burton, Oklahoma State University Cooperative Extension, Claremore, OK, 1985.

University of Vermont Extension. 2016. Fuel efficient lawns and landscapes. Website information available at http://pss.uvm.edu/ppp/articles/fuels.html; last accessed May 22, 2017.

Unruh, A.M., N. Smith & C. Scammell. 2000. The occupation of gardening in life-threatening illness: A qualitative pilot project. *Canadian Journal of Occupational Therapy, 67*(1), pp.70-77.

Vanhonacker, F., *et al.* 2010. How European consumers define the concept of traditional food: Evidence from a survey in six counties. *Agribusiness, 26*(4), pp.453-476.

References

WAHLQVIST AO, M.L. 2004. Requirements for healthy nutrition: integrating food sustainability, food variety, health. *Journal of food science, 69*(1), pp.CRH16-CRH18.

Warwick, H. 1999. Cuba's Organic Revolution. *Ecologist, 29*(8), pp.457-460.

Wilhelm, G. 1975. Dooryard gardens and gardening in the black community of Brushy, Texas. *Geographical Review, 65*(1), pp.73-92.

Windham, J.S. 2007. Putting Your Money Where Your Mouth Is: Perverse Food Subsidies, Social Responsibility and America's 2007 Farm Bill. *Environs: Environmental Law and Policy Journal, 31*, p.1.

Worldwatch Institute. 2016. Globetrotting food will travel farther than ever this Thanksgiving. Website information available at http://www.worldwatch.org/globetrotting-food-will-travel-farther-ever-thanksgiving; last accessed May 22, 2017.

Worsley, T., K. Baghurst & G. Skrzypiec. 1995. Meat consumption and young people. Final report to the Meat Research Corporation, CSIRO Division of Human Nutrition, Adelaide, SA, Australia.

Zahina-Ramos, J. 2013. Attitudes and Perspectives About Backyard Food Gardening: A Case Study in South Florida. Doctoral dissertation, Florida Atlantic University.

Zahina-Ramos, John. 2015. Just One Backyard: One Man's Search for Food Sustainability. Createspace Independent Publishing Platform.

Zakowska-Biemans, S. 2012. Traditional food from the consumers' vantage point. Zywnosc-Nauka Technologia Jakosc, 19, pp.5-18.

Zhang, Y., Y. Qian, D.J. Bremer and J.P. Kaye. 2013. Simulation of nitrous oxide emissions and estimation of global warming potential in turfgrass systems using the DAYCENT model. *Journal of environmental quality, 42*(4), pp.1100-1108.

About the Author

Dr. John Zahina-Ramos is the author of the award-winning book *Just One Backyard: One Man's Search for Food Sustainability*. He has written numerous technical documents and articles on the topics of water resources management, environmental modeling, ecological studies and environmental issues.

Dr. Zahina-Ramos has B.S and M.S. degrees in the Biological Sciences and a Ph.D. in Geosciences from Florida Atlantic University. His dissertation research focused on urban residents' attitudes and perspectives about urban agriculture. He holds a Certificate in Geographical Information Systems (GIS) from Florida Atlantic University and holds a Professional Wetland Scientist certification through the Society of Wetland Scientists. His professional career in the environmental sciences began in 1991 with Duke University's Nicholas School of the Environment, participating in Everglades ecological research that was focused on environmental impacts from agricultural runoff and water management activities. From 1997-2014 he worked for the South Florida Water Management District and managed projects that were focused on environmental restoration and sustainable consumptive water use.

Dr. Zahina-Ramos has taught adult education courses on landscaping and gardening, and college courses on environmental mapping, GIS and remote sensing (Palm Beach State College), environmental geology (Palm Beach State College), sustainable agriculture (Loyola University Chicago's Institute of Environmental Sustainability), environmental sustainability (Loyola University Chicago's Institute of Environmental Sustainability), and the scientific basis of environmental issues (Loyola University Chicago's Institute of Environmental Sustainability). He lives in the greater Chicagoland area with his husband and a very large backyard food garden.

www.ingramcontent.com/pod-product-compliance
Lightning Source LLC
Chambersburg PA
CBHW080455110426
42742CB00017B/2898